Thermal-Hydraulics for Space Power, Propulsion, and Thermal Management System Design

Edited by
William J. Krotiuk
General Electric Company
Astro-Space Division
Princeton, New Jersey

Volume 122
PROGRESS IN
ASTRONAUTICS AND AERONAUTICS

Martin Summerfield, Series Editor-in-Chief
Princeton Combustion Research Laboratories, Inc.
Monmouth Junction, New Jersey

Published by the American Institute of Aeronautics and Astronautics, Inc.
370 L'Enfant Promenade SW, Washington, DC 20024-2518

American Institute of Aeronautics and Astronautics, Inc.
Washington, DC

Library of Congress Cataloging in Publication Data

Thermal-hydraulics for space power, propulsion, and thermal
  management system design/edited by William J. Krotiuk.
  p.cm. – (Progress in astronautics and aeronautics;v.122)
  Includes bibliographical references.
  1. Space vehicles – Design and construction   2. Space vehicles –
Thermodynamics.   I. Krotiuk, William J.   II. Series.
TL507.P75   vol. 122   90–17910
[TL875]   629.1 ss – dc20   [629.47'4]
ISBN 0-930403-64-9

Table of Contents

Capillary Pumped Loop Supporting Technology 131

C. E. Braun, *General Electric Company, Princeton, New Jersey*

Chapter 3.   Startup Thaw Concept for the SP-100 Space Reactor
Power System 143

A. Kirpich, A. Das, H. Choe, E. McNamara, and D. Switick, *General Electric Company, Princeton, New Jersey*, and P. Bhandari, *Jet Propulsion Laboratory, Pasadena, California*

Preface

Until recently, thermal-hydraulic considerations have not been a major concern in spacecraft design. The power levels and heat dissipation requirements for past spacecraft did not warrant complex low-gravity fluid-thermal control systems. Because of these low power and dissipation requirements, sufficient margin could be built into a spacecraft's heat dissipation system to account for any low-gravity operational unknowns.

The larger weight, power, and heat dissipation requirements of planned spacecraft and structures make these past conservative design approaches inapplicable and undesirable. Some early space-based thermal dissipation systems employed one-phase fluid heat transfer loops. Many current designs employ heat pipes to enhance thermal dissipation. Systems employing two-phase heat removal loops are frequently being considered necessary for many future applications in order to reduce weight and achieve compactness. This design approach will become more important as space travel becomes more routine, making unique spacecraft designs less common and standardized designs more prevalent. Projects that exemplify these new space components and structures include the Space Station, the SP-100 Space Reactor, the Multimegawatt Space Nuclear Power Program, and NASA's Project Pathfinder.

Recent efforts have begun to expand the knowledge of fluid-thermal behavior for low-gravity space-based designs. This monograph represents one of the first attempts to bring together the technical knowledge necessary to design optimized fluid-thermal space systems. The technical scope of this volume is limited to the discussion of the latest thermal-hydraulic space designs needed to transfer large amounts of heat, and the phenomena associated with the behavior of one- and two-phase fluids in low gravity.

The technical discussions in this volume are divided into eight chapters. Chapters 1 to 3 discuss planned space-based projects, which employ low-gravity fluid-thermal systems and the specific problems encountered in their design. The remaining chapters are more theoretical and discuss recent advances in understanding the fluid-thermal phenomena needed to analyze and design low-gravity systems.

Chapter 1 briefly describes current and planned advanced space programs, especially their thermal-hydraulic aspects. *J. Alario* of Grumman Space Systems summarizes the fluid-thermal management systems being considered for the manned orbital Space Station. These systems consist of separated and mixed two-phase mechanical pumped loops, employing fluids such as ammonia or Freon. The proposed designs for the other important component of the Space Station fluid-thermal management system, the heat pipe radiator panels, are also described. *C. E. Braun*, formerly of General Electric Astro-Space Company, describes the proposed thermal management system for the Space Station free-flying platforms. A major component of the orbital free-flyers is a heat removal system employing a two-phase capillary pumped ammonia loop. The higher heat dissipation requirements specified for these proposed systems point to the need for two-phase flow cooling to achieve weight savings and compactness.

The nuclear power system design closest to prototype testing, the SP-100 Space Reactor Power System, is a one-phase liquid-metal cooled nuclear reactor. This design uses a thermoelectric power conversion method to create 100 kw of electricity for powering a spacecraft payload, station keeping and propulsion systems, and for lunar and planetary power applications. *A. Kirpich, G. Kruger, D. Matteo, and J. Stephen* of General Electric Astro-Space Company describe the design and operation of the SP-100 Reference Flight System. *J. A. Dearien and J. F. Whitbeck* of the Idaho National Engineering Laboratory discuss the Multimegawatt Space Power System concepts. These advanced designs are being proposed by various companies to develop larger levels of electrical power for orbital, interplanetary and planet surface applications. The designs emphasize the need for a better understanding of low-gravity fluid-thermal behavior.

Two-phase flow will also be important in the cooling of advanced propulsion systems components such as the power conditioning system and the high temperature nozzles, where the boiling process will allow efficient, compact cooling. The expulsion system can also possess two-phase flow conditions. *R. J. Cassady* of Rocket Research Company discusses these concerns.

Chapters 2 and 3 elaborate on the specific thermal-hydraulic concerns, experiments, analytical methods, and design solutions associated with two of the programs reviewed in Chapter 1. Chapter 2 specifically discusses the technology development, experimental plans, and analytical methods associated with the fluid-thermal management systems of the Space Station. *R. Brown and J. Alario* of Grumman Space Systems Company discuss mechanically pumped thermal bus systems and central heat-pipe radiator systems proposed for the Space Station, and *C. E. Braun* discusses the capillary pumped loop design of the orbiting platforms. The technology described here could ultimately be applied to other spacecraft designs.

In Chapter 3, *A. Kirpich, A. Das, H. Choe, E. McNamara, and D. Switick* of General Electric Astro-Space Company, and *P. Bhandari* of the Jet Propulsion Laboratories discuss a difficult problem currently being studied in the SP-100 Space Reactor. For safety reasons, space-based nuclear reactors will be launched cold and inoperable. Only in orbit would the reactor achieve criticality. Thus, the cooling system must be capable of reaching full operating capacity relatively quickly and assuredly. The SP-100 designers have chosen to cool the reactor with a liquid alkali metal forced-flow and heat-pipe system. The alkali metal will be launched frozen and will need to be thawed to reach full cooling capacity. The methods developed to thaw and start up the reactor system are described in this chapter.

Chapters 4 to 8 discuss the experiments, theories, and analytical methods associated with low-gravity fluid-thermal behavior. In Chapter 4, *Z. I. Antoniak* of Battelle, Pacific Northwest Laboratories and I describe the known low-gravity thermal-hydraulic experiments and trends. This chapter further discusses the numerical methods for performing low-gravity fluid-thermal calculations, and surveys and describes the available one- and three-dimensional computer programs. Past low-gravity experimental efforts the surveyed, and planned future experiments described. A description of the Earth- and space-based low-gravity experimental methods is also included.

An important aspect in modeling two-phase flow is the understanding and predicting of the flow distribution of the two phases in a channel. This phase distribution is called flow regime. Flow regime determination is important in predicting pressure drop and heat transfer. Chapter 5 contains a detailed description of recent low-gravity flow regime experiments. The author of this chapter, *A. E. Dukler* of the University of Houston, has developed methods for predicting flow regimes in normal-gravity two-phase pipe flow. In this chapter, he provides a theoretical description of a model to determine the appropriate low-gravity flow regimes. Initial trends in the calculation of low-gravity two-phase pressure drop are also described.

Boiling is an important means of removing heat from a surface, and condensation transfers the heat to another location for use or removal. Two types of boiling exist: pool (or low flow) and flow boiling. Boiling systems designed for space applications use forced convective boiling because this process possesses higher heat-transfer coefficients consistent with flow system operation. Low-gravity flow-boiling phenomena are discussed in Chapter 6. Observations and comparative calculational results from recent low-gravity flow-boiling experiments are described in this chapter by *J. Cuta* and me. *J. Cuta* of Battelle, Pacific Northwest Laboratories, also includes a discussion of the expected behavioral differences of various fluids during boiling. Special emphasis is placed on the behavior of cryogenic fluids undergoing rapid, unstable boiling. *Y. S. Chen* of SECA, Inc. describes a practical problem involving the calculation of boiling heat transfer at the ball bearings in the Space Shuttle high-pressure cryogenic oxygen turbopump. The calculation of the fluid, ball bearing and race temperatures are important to insure adequate operation of the Space Shuttle main engine and to improve future designs.

Chapter 7 discusses the low-gravity condensation phenomenon. *F. Best* of Texas A&M University provides a theoretical description of the expected drop, film and bulk condensation processes. Several low-gravity condensation experiments are described, including comparisons between the measured results and theoretical models. Dr. Best concludes by discussing the current state of the art and planned experiments.

Many people and organizations have contributed to the generation of this volume. I am indebted to Battelle, Pacific Northwest Laboratories, the U.S. Department of Energy, NASA, and the U.S. Air Force for supporting most of the work that was used as the basis for the discussions in this volume. Recognition is also due the General Electric Astro-Space Division for supplying support services necessary for completing the volume. Of course, I am indebted to the contributions of the various authors and the support of their employers. Their participation and cooperation made this effort possible. The encouragement and support of Dr. Martin Summerfield, the AIAA *Progress in Astronautics and Aeronautics* series Editor-in-Chief, during this project was also of great importance. I must also thank the AIAA for supporting the development of the volume and supplying the services of their staff, especially Mrs. Jeanne Godette, the series Managing Editor.

Finally, I would like to express my gratitude to three ladies who made my efforts in this activity possible; to RHK, who in the beginning supplied the encouragement to pursue the work that formed the basis of this volume, to CEGK, whose supplied inspiration was invaluable to the completion of this work, and to my daughter, Elise, whose simple, yet complex, thoughts are reflected in parts of Chapter 4. She is representative of the future generations who will ultimately be the recipients of the benefits of these activities.

William J. Krotiuk
March 1989

Chapter 1. Overview of Thermal-Hydraulic Aspects of Current Space Projects

Space Station Thermal Management Systems

Joseph Alario*
Grumman Space Systems, Bethpage, New York

Introduction and Background

Historically, the thermal management systems for spacecraft have been designed to be independent of gravity, performing the same on Earth as they do in orbit. The thermal management requirements of low-power satellites can usually be satisfied by passive devices such as coatings, heaters, and insulation. Early, manned missions such as Mercury, Gemini, and Apollo required the addition of a pumped fluid loop to collect greater amounts of waste heat from distributed sources for eventual rejection by space radiator or expendable coolant. These fluid systems were still essentially independent of gravity since heat transfer was dominated by forced convection. In addition, power consumption and long-term reliability were not major design drivers since pumping requirements were modest and mission life was relatively short. Even when power levels increased to 25 kW for the Space Shuttle, these conventional pumped fluid cooling systems were still appropriate for the mission requirements.

The Space Station changed that situation with requirements for much higher power levels and enhanced user utility. A constant-temperature heat-collection system (a thermal bus) was desired to permit standard user interfaces. In addition, the Space Station is being designed for a 30-yr mission life, low life cycle costs, and a flexible growth capability from an initial 75-kW system to a final 125-kW configuration. Crew safety and system reliability are dominant design drivers. The fixed configuration and single-point failure limitations of Shuttle-type pumped liquid loop radiators would not satisfy these requirements; a modular high-capacity heat pipe radiator would be preferable. Clearly, innovative approaches to thermal management were needed.

---

*Manager, Thermal Systems.

3

Under the leadership of Wil Ellis, Deputy Chief, Crew and Thermal Systems Division, NASA Johnson Space Center (JSC), a comprehensive plan was formulated at the Space Station Technology Workshop held at Williamsburg, Virginia in March 1983. Planning inputs were solicited from a broad base of government, industry, and university sources organized into a Thermal Management Working Group chaired by Robert Haslett of Grumman. The main thrust of the group's recommendations was to concentrate on the orderly development of breakthrough two-phase technologies so that they would be ready for early 1990's implementation on the space station. NASA accepted these recommendations and began a long-term development program by awarding a wide range of technology contracts in key areas that included high-capacity heat pipe radiators, two-phase heat-transport and -acquisition systems, maintainable heat-transfer interfaces, and high-performance evaporators and condensers. Each program followed a milestone approach, with incremental funding levels matched to demonstrated performance. In the typical development plan, innovative designs were first validated with proof-of-concept ground test articles. This was followed by more representative technology demonstration hardware for ground test validations and limited 0-g testing of critical components using NASA's KC-135 low-gravity test bed aircraft or, in special cases, the Space Shuttle. Testing with the KC-135 achieves a low, near-0-g field for 20–25 s during each of about 40 flight parabolas, whereas the Shuttle permits 0-g testing over much longer time periods.

The final phase of NASA's thermal technology development plan consisted of awarding parallel contracts for the design, fabrication, and testing of prototype thermal control systems for space station specific requirements. These various systems are in the process of undergoing competitive ground testing in NASA's Thermal Test Bed Facility at JSC. The preferred systems will be selected on technical merit after evaluation by NASA and prime contractor personnel. In addition to the prototype ground test hardware, two important Shuttle flight tests were included to demonstrate operational capability of prototype hardware in the space environment. The first Shuttle test, Space Station heat pipe advanced radiator element (SHARE), was originally scheduled for November 1985 as the final validation of the high-capacity monogroove heat pipe radiator. Development of this breakthrough technology began in 1979, in which an order-of-magnitude improvement in heat pipe performance was sought that would satisfy space station requirements. Unfortunately, the Challenger disaster on flight 51L forced an extended delay until March 13, 1989 when SHARE was flown on STS-29. Detailed evaluation of the flight data is still ongoing and has revealed problems with manifold priming and bubble management in 0g that prevented proper startup and operation of this full-size test article. Design modifications have been developed to ensure comparable performance under both 1-g and 0-g conditions. These changes are now being validated in a series of 0-g experiments for the KC-135 aircraft and the shuttle mid-deck. Another full-size radiator test article, SHARE II scheduled for January 1991, will confirm the fixes. This will be followed by the second type of planned Shuttle test, two-phase integrated thermal system/shuttle radiator assembly demonstration (TPITS/SRAD), which combines evaluation of space erectable radiator assembly/maintenance procedures and interface devices and validation of a two-phase heat-transport/acquisition system (i.e., two-phase thermal bus). The TPITS/SRAD experiment is scheduled for August 1992.

Active Thermal Management System Descriptions

The space station active thermal management systems can be grouped into three broad categories: the central active thermal control system (ATCS), the electrical power system TCS, and the payload/platform TCS. Each application has performance requirements that could result in different thermal control approaches, choosing between various combinations of pumped two-phase loops, capillary pumped loops, and high-capacity heat pipe radiators.

The role of the central ATCS is to provide a constant-temperature interface for a wide variety of distributed heat sources. This is illustrated in Fig. 1. Two different temperature levels are provided; one at 35°F to service metabolic heat loads and one at 70°F for equipment cooling loads. Each temperature level uses redundant loops for increased reliability and survivability. Table 1 summarizes the ATCS design requirements.

The main elements of the central ATCS are the heat-transport and -acquisition system and the central heat-rejection system. The heat-transport and -acquisition system, commonly referred to as the thermal bus, relies on the principle of two-phase heat transfer (evaporation at the heat sources and condensation at the heat sinks) to provide a near-constant loop temperature with reduced pumping power requirements.

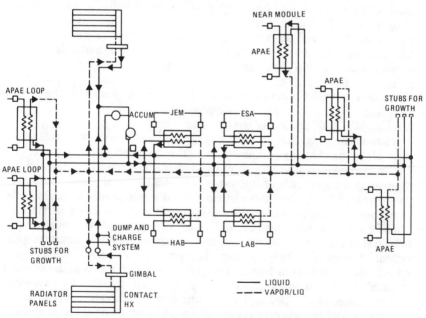

Fig. 1   Space Station thermal management schematic.

Table 1 ATCS design requirements

• Total heat load:	Initial operating configuration = 70 kW (growth = 125 kW) 20 kW at 2°C ± 2.5°C 50 kW at 21°C ± 10°C
• Heat load turndown ratio:	$\dfrac{\text{Maximum heat load}}{\text{Minimum heat load}} = 10/1$
• Min radiator temp:	−80°F
• 10 year life with 99% reliability for system	
• Min radiator area (η (efficiency) > 90%)	

Previously, the heat-transport systems used for manned space flight have utilized pumped single-phase fluids such as water, water-glycol, and Freon. They have performed well on programs such as Mercury, Apollo, and the Space Shuttle. However, the greatly increased heat load requirements of Space Station Freedom and similar future large space platforms make it prohibitively more difficult to use these same conventional sensible heat-transport systems to maintain the closely regulated coolant temperatures needed for both metabolic and equipment thermal control. The obvious performance benefit of two-phase heat-transfer/transport vs the conventional pumped liquid loop systems, in terms of required flow rate and hence weight and power penalties, is shown in Fig. 2. Ammonia is the preferred fluid for all space station heat-transport systems located in any unpressurized/unmanned environment where toxicity is not a safety issue. The more important characteristics of an optimum fluid for two-phase systems include the following: compatibility with common construction materials, moderate vapor pressure within the desired operating temperature range, low freezing point, high latent heat of vaporization, high density, and low viscosity. As seen, the constant-temperature ammonia two-phase system requires much less fluid circulation than a single-phase system that is constrained to a comparable small allowable temperature rise.

It should be noted that pumped single-phase water loops must still be used in all of the manned environments due to the stringent toxicity requirements. Two-phase systems that use water as the working fluid are not practical in the temperature regimes of interest on the space station (35–100°F) due to the correspondingly low vapor pressure and density, which would dictate excessively large line diameters to keep vapor pressure losses very small.

There are two fundamentally different two-phase heat-transport/acquisition systems under development for the space station: separated-phase systems and mixed-phase systems. The basic schematics of each type of system are illustrated in Fig. 3. The generic components that are common to each system are evaporators (heat supply), condensers (heat removal), and a pump (fluid circulation). Radiators are shown coupled to the condensers for ultimate heat rejection to space. Two common terms are used in conjunction with two-phase systems: system set-point temperature, which

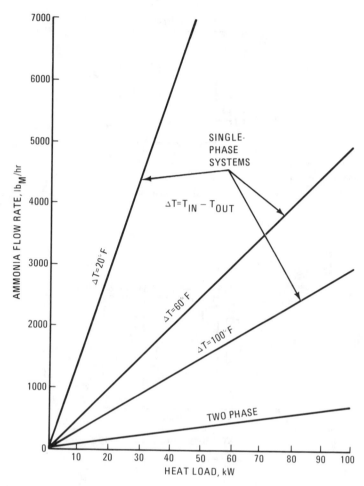

Fig. 2   Flow requirement for single- and two-phase systems.

is the vapor temperature immediately downstream of the evaporators, and degrees of subcooling, which is the temperature difference between the pump inlet (liquid) and system set point (saturated vapor). In the separated-phase concept the liquid supply to the heat sources is precisely controlled to match the locally applied heat load. This prevents carryover of excess liquid and results in a very-high-quality vapor being returned to the condensers. The primary advantage of the separated-phase concept is well-understood conventional forced-flow liquid and vapor fluid dynamics, which minimizes ground testing problems and permits more accurate design of a full-size system for 0-g operation. Also, fluid circulation is provided by a single conventional mechanical pump. The main disadvantage is in providing a reliable method of controlling liquid flow to the evaporators.

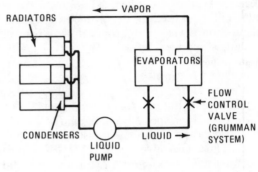

a) Separated

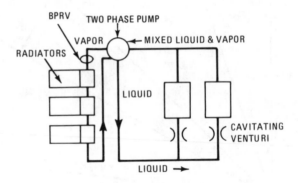

b) Mixed, two-phase pump

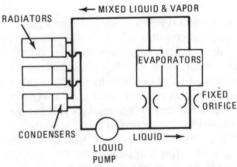

c) Mixed, liquid pump

Fig. 3 Schematics of separated and mixed two-phase systems.

Separated-phase systems are under development by two NASA contractors, Grumman and Lockheed. In the Grumman suystem, the liquid flow to each heat acquisition device (i.e., evaporator) is controlled by a capillary feed system that is supplied by a local reservoir. The liquid within the reservoir is maintained by the action of a solenoid control valve whose on/off operation is triggered by an ultrasonic measurement of the liquid volume. The Lockheed evaporator also relies on a capillary wick system to provide a matched liquid supply, but in a different manner. Excess liquid is pumped to the wick, which has been specially configured to draw only what is needed to meet the heat load and return the unused liquid to the condenser via a condensate return line. The valve/sensor control system has been eliminated, but a more complicated wick and an extra plumbing line are required.

In the mixed-phase concept, a constant liquid flow is always supplied to the evaporators, which results in a mixed liquid/vapor flow of widely varying quality in the return line. The mixed-phase, constant liquid flow concept simplifies the evaporator design of the separated-phase systems but results in a difficult analytical/test correlation problem associated with two-phase flow in the vapor line. Two-phase flow is not well understood in 1 g and has received limited testing in 0 g only recently, with much work remaining to correlate theory with practice[1] (see also Chapters 4 and 5). This initial microgravity test data indicate an early transition to annular flow (at 15% quality) and continuing annular flow through high quality (at 90%) with a theoretical transition to mist flow at higher quality (> 90%). The two-phase pressure drop is strongly affected by flow regime. In microgravity, two-phase pressure drop is significantly greater than that under 1-g conditions at the same conditions of mass flux and quality.

Two viable mixed-phase thermal bus systems are being considered for the Space Station. One is under development for NASA by Boeing and features three unique components (Fig. 3b); a cavitating venturi at each evaporator inlet, a two-phase pitot-type pump called a rotary fluid management device (RFMD), and a backpressure regulating valve (BPRV). The cavitating venturi is used to maintain constant inlet flow rate independently of the downstream outlet pressure; the RFMD relies on centrifugal force within a spinning drum to both separate ammonia liquid from vapor and increase the pressure within the liquid; and the BPRV is used to regulate the condenser pressure in response to operating conditions.

The other mixed two-phase concept has been developed at Grumman and demonstrated in recent thermal test bed evaluations at NASA-JSC. It uses all of the basic components of the Grumman separated-phase concept except for the evaporator control valves, which are replaced by fixed orifices. Thus, mixed liquid and vapor is allowed to enter the vapor return line and flow to the condensers where the mixture condenses completely and becomes subcooled liquid. The liquid circulation is accomplished with a conventional pump for simple, more reliable operation.

In any of these candidate systems, maintaining a stable set-point temperature over a wide range of heat load is accomplished by matching the heat rejected at the condensers to the total heat load applied to the evaporators. Any one of three basic techniques can be used to control condenser heat

Table 2 Comparison of distinguishing features, thermal bus systems

	Separated phase	Mixed phase
• Boeing		• RFMD pitot pump
		• Bellows accumulator (liquid/vapor)
		• BPRV
		• Cavitating venturis at evaporators
• Grumman	• Conventional pump	• Conventional pump
	• 2ϕ Accumulator	• 2ϕ Accumulator
	• Condenser liquid flooding	• Condenser liquid flooding
	• Capillary feed of evaporators from local reservoir	• Flow-through evaporators
	• Solenoid control valve at evaporator/ reservoir	• Fixed orifice at evaporator
• Lockheed	• Conventional pump	
	• Bellows accumulator (liquid/gas)	
	• Condenser gas blockage	
	• Fine capillary wicks (1 μ) in evaporator	
	• Recirculating liquid line from each evaporator	

rejection: liquid blockage, gas blockage, or condenser pressure (and temperature) regulation. The choice depends on what is deemed best for the system being considered. Liquid blockage (condenser flooding) is used in the Grumman systems and relies on the control of excess liquid (normally held in an accumulator) to block or unblock the condensation area required to match the evaporation load. Gas blockage, which is used in the Lockheed bus, accomplishes the same spoiling of the condenser area by the action of a noncondensable inert gas such as nitrogen. It is easier to test since the gas behavior in a 1-g ground test environment closely simulates the 0-g conditions. Controlling the condenser temperature by regulating its backpressure is the technique used by Boeing in its mixed-phase concept, although a certain amount of condenser flooding also results. A listing of the distinguishing features of each thermal bus system under consideration is given in Table 2.

Heat Pipe Radiator Panel Description

In all cases, the heat released within the thermal bus condenser is transferred to high-capacity heat pipe radiator panels that are attached to the outside surfaces of the condenser heat exchangers. A dry contact interface is maintained between the evaporator sections of the radiator panels and the bus condenser shells. Maintainable clamping mechanisms are used to apply the required pressure needed to achieve the desired minimum heat conductance at the dry interface (500 Btu/h-ft^2-°F). One attachment/interface concept is illustrated in Fig. 4, where small interconnected beam elements are used to uniformly distribute a centrally applied tension load. Details are provided in Chapter 2. It is intended to permit easy initial on-orbit installation and subsequent maintenance and replacement of damaged panels. Replacement and servicing in this manner is the only practical way of achieving the high mission reliability and long life without prohibitive weight and cost penalties.

An alternate method of clamping the radiator to the thermal bus is the Lockheed/LTV pressurized bladder, with suitable honeycomb backup structure. The pressurant used can be either inert nitrogen gas or hydraulic fluid. Additional information is also contained in Chapter 2.

Early system requirement studies set the desired heat pipe panel length at approximately 50 ft to facilitate economical accommodation within the Shuttle Orbiter cargo bay. A minimum capacity of 1.5–2 kW was established to enable all of the waste heat from the initial 75-kW Space

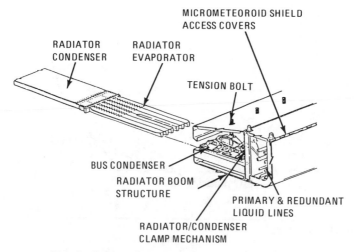

Fig. 4 Radiator/condenser dry contact interface.

Station to be rejected with about 50 heat pipes, while still leaving ample performance margin for each panel. This approach minimizes the number of individual panels required by the radiator system, but still provides an overall system that is relatively insensitive to the loss of any single radiator element. As seen in Fig. 5, a redundancy level of under 10% and scheduled maintenance is typically required to meet system performance requirements over the entire mission life. However, to achieve this ambitious design goal for the heat-rejection system, both the heat-transfer and heat-transport requirements that were being imposed on these high-capacity heat pipe radiator panels demanded an order-of-magnitude increase in the existing heat pipe state of the art. Clearly, a technology breakthrough was needed. This occurred in the early 1980's with the innovation of the monogroove heat pipe concept under the sponsorship of NASA-JSC.[2]

The two candidates' designs now under development for the high-capacity heat pipe radiator panels are both related to the monogroove heat pipe concept (Fig. 6). In the monogroove concept, the countercurrent liquid and vapor flow are each isolated in separate channels that are joined only by a very narrow rectangular slot. The slot carries the required flow of liquid to and from the liquid channel at the condenser and evaporator regions and also supports the capillary pressure differences that exist between the channels. Fine wall grooves are machined in the upper (vapor) channel to feed liquid from the liquid channel for evaporation in the heat

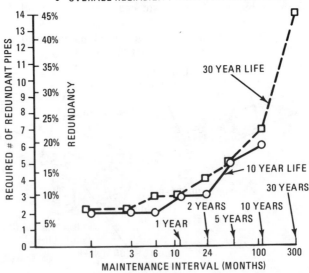

Fig. 5 Sensitivity to maintenance interval and lifetime.

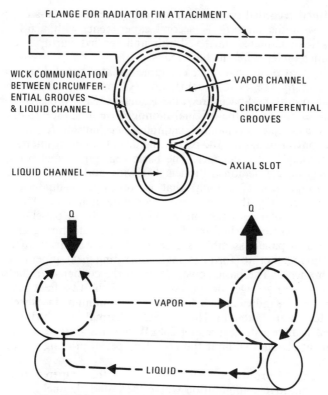

Fig. 6 **Monogroove heat pipe operating principle.**

input zone and to return condensate to the liquid channel in the condenser section (heat-removal zone).

This design exemplifies the breakthrough approach needed for the next generation thermal management systems. The configuration does not penalize on-orbit performance, yet still meets the requirement of ground operation for total system checkout. In the case of the monogroove heat pipe, most operating characteristics can be measured in ground tests except for the initial filling of the liquid channel, which, on the ground, is assisted by gravity because of its relatively large diameter. Complete validation of the concept required special experimentation to confirm proper filling and operation of the liquid channel in a 0-*g* environment. Once the basic 0-*g* operating principles are confirmed, then ground tests can be used with confidence for other performance measurements during development and production of prototype hardware.

The monogroove "basic principle" concerns were answered early in NASA's development program by two relatively low-cost tests. The first was on NASA's low-gravity KC-135 aircraft test bed. Short lengths of the heat pipes were cut lengthwise and resealed with a glass facesheet. Liquid channel priming from many different initial liquid puddle orientations was observed and recorded on movie film during the 20- to 30-s 0-g parabolas available with the aircraft. A second experiment, with a 6-ft heat pipe radiator, was conducted on the Shuttle and verified startup and operating performance under different heat loads.

Two different prototype space erectable radiator panel designs are under development by the NASA contractors Grumman and Lockheed. The Grumman prototype Space Erectable Radiator (SERS) design is based on a low-risk extension of the original monogroove heat pipe technology and also uses a simple mechanical clamping mechanism for achieving the required contact pressure (100 psi) at the dry attachment interface with the bus condensing heat exchanger. The basic heat pipe panel being built for ground test article demonstration is 1 ft wide and consists of a compact multileg evaporator (2 ft long) that is joined to a dual-leg, 46-ft-long condenser section both welded to a common manifold. The Grumman design uses modular construction in which the radiator panel is made up of separate subassemblies that are designed for manufacturing and handling ease. The final panel assembly is comprised of six separate subassemblies, which join together to form a single operating heat pipe with multiple parallel branches. The basic construction of the condenser/radiator fin is a simple sheet and stringer design consisting of the two heat pipe legs and edge channels sandwiched between two aluminum facesheets. Recent ground test data taken at NASA-JSC's thermal test bed facility have confirmed heat transfer exceeding 2.4 kW and response to rapid load and environment changes without the need for special priming or conditioning procedures.

The Lockheed prototype SERS design uses their tapered artery version of the monogroove configuration where the liquid channel is shaped like a teardrop. Lockheed believes that this shape improves the fluid dynamic stability at the interface between the capillary slot meniscus and the circumferential wall grooves.

Lockheed's SERS panel incorporates two separate heat pipes, each with a single condenser leg and three evaporator legs. These heat pipes are first manufactured as separate complete assemblies; then they are bonded with a panel that is made from aluminum honeycomb core and facesheets. Thus, the heat pipe segments are joined together during the initial heat pipe fabrication step and not as part of the final panel assembly operation, as is the case with Grumman's approach. Lockheed actually supplies the heat pipes to a subcontractor, LTV, which is responsible for making the finished radiator panel.

Additional technical details for both the thermal bus systems and the radiator systems are provided in the chapter by Brown and Alario.

References

[1]Chen, I., Downing, R., Keshock, E., and Al-Sharif, M., "An Experiment Study and Prediction of a Two-Phase Pressure Drop in Microgravity," AIAA Paper 89-0074, Jan. 1989.

[2]Alario, J., Haslett, R., and Kosson, R., "The Monogroove High Performance Heat Pipe," AIAA Paper 81-1156, June 1981.

Thermal Design of the Space Station Free-Flying Platforms

C. E. Braun*
General Electric Company, Princeton, New Jersey

Platform Thermal Design Overview

The current Space Station system consists of a manned main station complex complemented by unmanned Earth-orbiting platforms. The unmanned platforms, which are part of NASA/GSFC work package 03, will influence the design of future government and commercial spacecraft. Cost savings and other benefits can be realized by developing generic subsystems and interfaces that can be used on a wide variety of such platforms. Commonality will simplify operation, maintenance, and expansion of space platforms and allow changes in requirements over an operational life of 10–20 yr or more. Thermal management subsystems have been identified by NASA as especially critical to the design of these platforms.

Space station platforms offer a wide range of mission capability, from polar-orbiting Earth observation and research to celestial viewing telescopes. In contrast to a typical mission life expectancy of 5 yr for low-Earth-orbiting (LEO) spacecraft, the mission life of a space platform will exceed 20 yr. Extended lifetimes became viable by means of in-orbit maintenance and replacement of limited-life subsystem and payload components. Modular subsystem design facilitates such replacement and allows an increased resource capability (i.e., power, data rate, heat rejection) to satisfy demands for future missions.

Thus, the thermal control subsystem (TCS) on unmanned free-flying platforms must fulfill more stringent design requirements than those of previous spacecraft. Provisions for high heat fluxes, precise temperature

*Manager, Thermal Systems Technology, Astro-Space Division; currently Manager, Thermodynamics, Fairchild Space Company, Germantown, Maryland.

control, and system serviceability/modularity present new challenges to the thermal designer. A series of trade studies led to a "hybrid" TCS design, in which a central core fluid loop is used for thermal control of the payload experiments, whereas the subsystems are independently regulated. Platform system requirements of high reliability and weight optimization favored the use of a capillary pumped central fluid loop. Ease of integration with the platform structure was also a TCS design consideration.[1]

The thermal management design approach for the Space Station unmanned free-flying platforms accommodates system-level modularity with on-orbit replaceable radiators, cold plates, and control hardware. Basic to the TCS design is its ability to perform without the need for thermal system redesign in numerous orbital environments with payload (PL) sets varying in power level and configuration.

Platform Thermal Control Requirements

The polar platform flies in a 98.5-deg inclination with operating altitude ranging from 500 to 900 km. Required nodal crossing times of 1300 and 0830 h result in an effective sun angle range (angle between sun vector and orbital plane) between 9 and 55 deg (Fig. 1). The platform's $-Y$ axis points toward cold space.

Co-orbiting missions fly in a 28.5-deg inclination with operating altitude ranging from 460 to 1000 km. Co-orbiting telescope missions typically have strict sun-avoidance criteria, requiring the platform to perform a 180-deg rollover maneuver to prevent the sun from entering the telescope's aperture. Half of the telescope's circumferential body continually faces direct sun, whereas the opposing half continually faces cold space (Fig. 2).

Platform design features were derived following a review of the NASA-Langley Data Base (MRDB) and data gathered from the NASA Customer Integration Office. As shown in Fig. 3, polar-orbiting platform (POP) missions initially comprise 12–20 payload experiments with a potential growth to 30 experiments. Thermal dissipation of each payload, as required in the NASA polar payload accommodation test set (PPATS), ranges from 10 to 600 W with a mission set dissipation totaling 3 kW. A 20–25°C temperature control range at the payload-to-platform thermal interface accommodates virtually all of the instruments in the database.

Co-orbiting platforms (COP) (Fig. 2) for the co-orbiter payload accommodation test set (CPATS) support cryogenically cooled infrared (IR) telescopes that are equipped with a sophisticated thermal control system and typically require thermal isolation at the payload-to-platform interface.

Platform subsystems such as power, communications, and data management are housed in orbital replacement units (ORU's). In general, common subsystem ORU design is desired for all platform (POP and COP) missions. To prolong equipment life, the ORU TCS is required to maintain component interfaces between 5 and 25°C.

Through Space Station Phase B studies, NASA Systems Engineering and Integration (SE&I) has developed thermal system requirements; critical

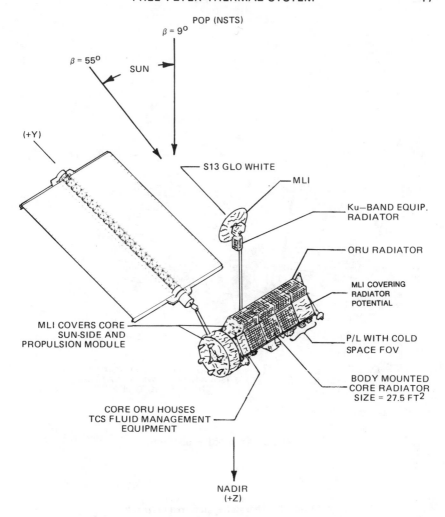

Fig. 1 Polar orbiting platform.

requirements from NASA's contract end item (CEI) specification are summarized in Table 1.

Mechanical Configuration Overview

The platform mechanical configuration consists of a carrier (core) structure, ORU structure, payload assembly plates, and secondary structure. The 23×10 ft carrier core supports Shuttle-launched polar mission sets, whereas a longer 33×10 ft carrier is used for Titan-launched polar missions. Housekeeping components are packaged in standard ORU housings $28.5 \times 28.5 \times 65$ in. POP ORU's are typically mounted on the $(-Z)$ zenith

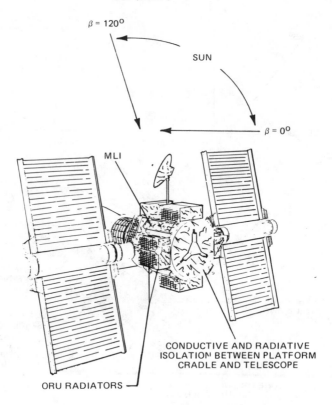

Fig. 2   Co-orbiting platform.

Table 1   Thermal system requirements

	Heat rejection, W		
	Platform TCS[a]	Payload TCS	Total
POP STS	1,050	350	1,400
POP ELV	2,400	800	3,200

Temperature control (nominal operation)

Payload interface (POP)	20–25°C
General housekeeping	5–25°C
Battery	−5–10°C

[a]75% of total payload heat load.

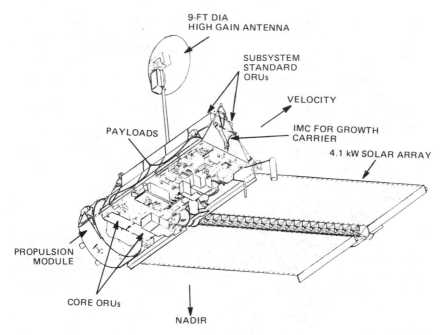

Fig. 3 Polar orbiting platform.

carrier side, whereas payloads mount on the (+ Z) nadir face. A cradle or ring-type structure wraps ORU's around the larger telescope structure of co-orbiting missions.

Payload assembly plates are the standard interface between POP payloads and platform core structure. Payloads are integrated on the ground to the assembly plate and then connected to the platform either before initial launch or during an on-orbiting service mission. A standard interface connector (SIC) provides a means of mechanical support, electrical power, data transmission, and thermal transfer between the platform and payload assembly plate. Platform growth is accommodated via the addition of a carrier structure containing the subsystem ORU's that support the growth payload. The growth carrier also contains all of the necessary thermal hardware (radiators, cold plates, fluid lines) needed to control payload temperatures.

Thermal Control Subsystem Description

The TCS provides all functions associated with temperature control of payload interfaces and spacecraft system components. Hybrid TCS architecture provides centralized heat collection and rejection for POP payloads and distributed controls for platform system ORU's. The central thermal bus consists of body-mounted radiators (BMR's), cold plates, fluid lines and a reservoir, quick disconnects, and isolation valves. Each ORU is a self-contained (or distributed) thermal system permitting temperature control independent of the central bus. Radiators supporting payload heat rejection envelop the carrier structure, and radiators rejecting housekeeping thermal loads surround the ORU structure. Temperature sensors, pressure transducers, and heater control hardware maintain autonomous functioning of the TCS during nominal operation. Continuous monitoring of the TCS by the applications software package enhances TCS performance. Software functions include telemetry status and limit checking, trending, system startup, heater set-point control, load management, leak detection, and system isolation control.

Thermal system redundancy ensures full operational capability after a single failure and provides safe-mode capability (i.e., maintains operation of critical housekeeping functions) after a second failure.

Trade Studies

The baseline platform configuration evolved from system trades maximizing payload mass to orbit with a maintainable spacecraft that can accommodate growth at a minimum system cost. TCS trades converged on solutions that provided the required level of performance and flexibility with lightweight, low-risk components. Major trade categories included TCS architecture, radiator configuration, and heat-transport technology.

The TCS architecture is a selection among the following:

1) a distributed system in which the heat dissipation requirements for the PL and housekeeping subsystems are handled locally through discrete radiating surfaces;

2) a centralized heat-collection/transport system connecting the PL and housekeeping ORU's to radiating surfaces; or

3) a hybrid system in which an optimized combination of the central and distributed systems has been tailored to the overall thermal control requirements.

Distributed System

Distributed TCS architecture offers thermal hardware simplicity and reliability at the expense of system integration and configuration. Thermal radiators, local to the PL or subsystem ORU's, dissipate power at optimum temperature levels. Heat transfer across mechanical interfaces is not required for system modularity. Flight-proven heat pipe technology can be used for local heat spreading.

Development of a configuration in which the many distributed (local) radiators have an appropriate field of view (FOV) to a cold sink poses a design challenge. Detailed understanding of payload thermal design and accurate modeling of interaction among the payload radiators of a mission set is required. Payload radiator FOV requirements can dictate different payload carrier configurations for the various mission sets. The fully distributed architecture was rejected due to unacceptable configuration requirements and lack of generic thermal design.

Centralized System

The centralized TCS approach offers configuration flexibility with relatively high system complexity and cost. Equipping potential PL and ORU mounting sites with the appropriate interface to the thermal bus eliminates thermal constraints imposed on instrument/ORU layout. Tailoring a cold plate to meet special PL or ORU power-handling requirements eliminates thermal packaging constraints. In general, centralized architecture provides a generic thermal design solution, independent of the platform, mechanical configuration. To meet Space Station requirements, however, a complex TCS results. Satisfying system reliability temperature levels could require as many as five thermal loops: 1) primary 20°C loop; 2) backup 20°C; 3) fail-safe loop for critical housekeeping; 4) primary 8°C battery loop; and 5) backup 8°C battery loop.

In addition to the 3-kW payload heat-rejection requirement, the platform central radiator must reject housekeeping thermal loads of up to 2.3 kW. The baseline platform configuration cannot handle this load with a body-mounted radiator, and therefore the complexities and penalties (see next section) associated with a free-standing radiator (FSR) design would be imposed on the system. Payloads of the co-orbiting telescope mission are equipped with their own thermal control system and require isolation at the platform interface. Centralized ORU heat rejection would impose the requirement for a central bus (including FSR) on the COP mission solely for control of housekeeping equipment.

Hybrid System

The choice for the POP architecture is an optimized combination of the central and distributed systems. A central two-phase bus is employed for the PL set, whereas a distributed radiator approach is used for the housekeeping ORU's and those PL locations having an adequate cold-space FOV. The central bus configuration permits integration of payload sets on the platform without special regard for payload layout (e.g., cold space for payload radiator). Judicious use of distributed thermal control takes advantage of the available radiating surface applicable to the POP and COP. Platform orbital views and operating environments for POP and COP dictate the positioning of specific ORU's on the spacecraft. Special housekeeping thermal control requirements, such as the "power" and "high-rate tape recorder" ORU's, are accommodated by mounting locations on the cold side of the platform. Overall, the hybrid architecture derives maximum benefit from both the central and distributed approaches.

Radiator

Selection of a POP radiator configuration resulted from trades between an appendage-type FSR and the integral BMR. The FSR configuration is area efficient if a radiator gimbal mechanism is used to optimize cold-space view. In addition, heat-rejection capacity is independent of platform carrier size. However, the FSR requires space construction or deployment and is a potential obstruction of FOV's of other platform appendages [e.g., solar array and tracking and data relay satellite system (TDRSS) antenna]. The FSR can also contribute to platform disturbance torque. The BMR approach offers distinct assembly and integration advantages. The platform design is compatible with an expendable launch vehicle. The BMR launched in its operational flight configuration eliminates developmental costs associated with an FSR deployment mechanism. Low risk and negligible impact on other platform subsystems were the primary drivers for choosing a BMR configuration.

Heat Transport for Central Bus

Heat-transport requirements for the thermal bus were derived using payload dissipation levels, transport distance between the cold plate and radiator, and the required temperature control range. Candidate technologies included high-capacity heat pipes, a single-phase fluid loop, a capillary pumped two-phase fluid loop (CPL) system, and a mechanically pumped two-phase fluid system (MPS). Heat pipes, which are closed-loop heat-transport segments, lead to large temperature gradients and interface weight penalties between the heat pipe segments when used with a modular bus design. A single-phase fluid system, with the temperature gradients inherent in a system relying on sensible heating, is not a weight/power optimized solution for the $22 + 2.5°C$ cold-plate interface temperature requirement. Both the CPL and two-phase MPS are easily adaptable to thermal bus modularity (i.e., quick fluid disconnect) and meet the relatively low heat load (3 kW) and short transport distance (less than 50 ft) requirements of the platform carrier TCS. The inherent simplicity of passive capillary fluid regulation and pumping, eliminating the active flow control associated with mechanically pumped systems, was the primary criteria for choosing a capillary pumped thermal bus as the baseline TCS.

ORU Thermal Control

The standard platform ORU is equipped with a stand-alone or self-contained thermal control system. Two external surfaces of the ORU structure (each 12.9 ft^2) are used as thermal radiators. These perpendicular radiator surfaces view multiple space sinks and provide useful heat rejection for all proposed platform, POP, and COP mounting locations (see Fig. 4). In general, high heat-dissipating components, such as the 1.25-kW power ORU's, require favorable cold-side mounting, whereas low-powered ORU's can easily survive in the warmer, sun-side environment. ORU radiator rejection capacity, provided by the proposed platform configurations, support the stringent component temperature levels (5–25°C) required by the Space Station Program.

To minimize temperature rise and weight, high heat-dissipating components mount directly to the radiator surface. As shown in Fig. 5, the two ORU radiator surfaces are thermally coupled by variable-conductance heat pipes (VCHP's). The surface to which a component mounts is considered the fixed radiator (i.e., rejects heat independent of VCHP operation). During beginning-of-life (BOL) or operation in cold environments, the noncondensable gas within the VCHP reservoir expends and shuts down the variable radiator segment, thus minimizing heater power. Heat pipe

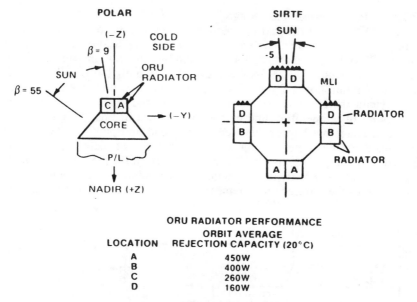

ORU RADIATOR PERFORMANCE
ORBIT AVERAGE

LOCATION	REJECTION CAPACITY (20°C)
A	450W
B	400W
C	260W
D	160W

Fig. 4   ORU radiator placement.

transport capacity and redundancy layout is such that failure of one pipe will not degrade radiator performance.

Component heat load was a factor in the ORU packaging studies. In general, ORU radiator power density yields a component baseplate temperature below the 25°C Space Station requirement. Because of the special temperature requirement of batteries, the power ORU radiator is segmented between the power management and distribution (PMAD) and battery equipment. The PMAD radiator interface operates at the nominal 25°C, whereas the battery radiator is sized to provide a nominal cell temperature below 10°C. Conductive mounting sleeves maintain a less than 3°C gradient between and on the case of battery cells.

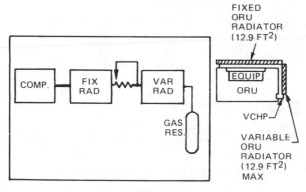

Fig. 5   ORU subsystem.

POP Payload Thermal Bus Design

The POP payload thermal bus performs the functions of heat collection, transport, and rejection as well as providing make-up heat during quiescent modes. Two loops, each capable of handling the full instrument load, provide the required TCS redundancy. Should both loops fail, survival heaters maintain nonoperational temperature limits. The block diagram of the payload thermal bus and its relationship to the loop's thermodynamic state point of temperature and entropy are shown in Figs. 6 and 7.

The interface mechanism between the payload and platform TCS is the cold plate. As illustrated in Fig. 6, although the cold plate is part of the payload assembly plate, its physical separation from the mechanical mounting plate allows flexibility in cold-plate design and integration. Tailoring the cold plate to the interface requirements of the NASA PPATS evolved two

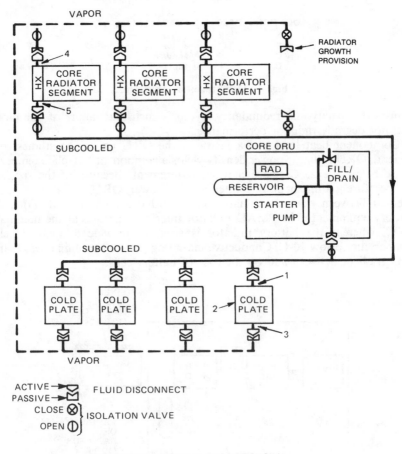

Fig. 6 Payload thermal bus.

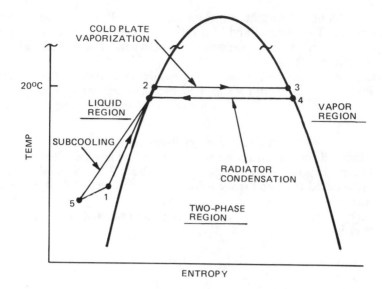

1 TO 2 LIQUID SENSIBLE HEATING WITH PRESSURE RISE (CAPILLARY PUMP)

2 TO 3 EVAPORATION IN CAPILLARY PUMP

3 TO 4 SATURATED VAPOR FLOW WITH PRESSURE DROP

4 TO 5 CONDENSATION AND SUBCOOLING WITH PRESSURE DROP

5 TO 1 LIQUID FLOW WITH PRESSURE DROP

Fig. 7   Payload thermal bus thermodynamic conditions.

generic sizes: 2.5×2.5 ft and 5.0×2.5 ft. The cold plate is the key contributor to the TCS weight. Limiting power-handling capability to 500 and 1000 W, respectively, allowed use of lightweight capillary pump extrusion (i.e., HS-111) and achieved cold-plate specific weights of less than 2.0 lb/ft^2.

Each cold plate contains redundant capillary pumps for heat collection, fluid pumping, and regulation. Subcooled liquid entering the cold plate is vaporized at the system saturation temperature. During evaporation, capillary action within the cold plates establishes a pumping pressure of 0.5 psid. The pumping action of this process is proportional to the heating load and as a result, saturated vapor exits the cold plate without the need for active flow control valving. Fluid quick disconnects in the SIC couple the cold-plate fluid system to the thermal bus.

Vapor is transported from the cold plate through smooth-wall tubing to a redundant low-pressure-drop heat exchanger element. The 27.5-ft^2 BMR segment, mounted to the cold ($-Y$) carrier side, has a minimum heat-rejection capacity of 500 W. Radiator segments are integrated in parallel to minimize pressure drop and are sized such that fluid is subcooled when it leaves the condensing heat exchanger. Heat pipe capacity and redundancy layout are designed so that failure of one pipe will not degrade radiator performance. The radiator and heat exchanger segment are separable from the platform core by fluid, mechanical, and electrical disconnects within the radiator spacecraft.

System saturation temperature and pressure are maintained by a heater-controlled fluid reservoir packaged in the disconnectable core ORU. A capillary starter pump is used for efficient cold-plate priming.

Thermal System Performance

The thermal control subsystem was evaluated by analyzing the polar and co-orbiting missions of the NASA-PATS. The TRASYS[2] models of the Shuttle-launched POP and space infrared telescope facility COP mission is shown in Figs. 8 and 9. A worst-case combination of orbital parameters, equipment operating mode, and thermophysical properties was assumed for the analysis.

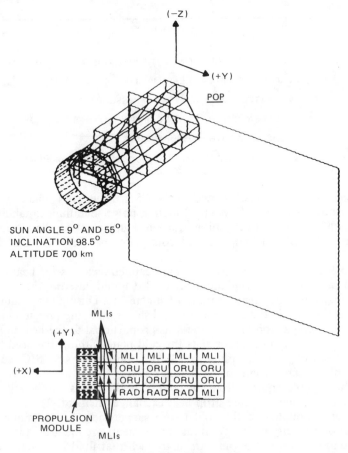

Fig. 8 POP TRASYS model.

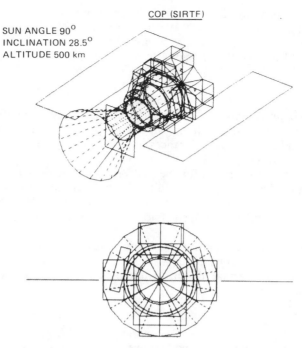

SUN ANGLE 90°
INCLINATION 28.5°
ALTITUDE 500 km

COP (SIRTF)

Fig. 9   COP (SIRTF) TRASYS model.

The Shuttle POP configuration can carry four core radiator segments totaling 110 ft² and dissipating up to 2000 W. Therefore, only three radiator segments are needed to provide the required cooling capacity (1.4 kW, avg; 2 kW, peak) for the STS-PPATS mission. The increased core length of the ELV-launched platform configuration provides an additional 55 ft² of radiator area and increases heat-rejection capacity to 3 kW with six radiator segments.

Thermal bus components are designed for the low pressure drop requirement of a capillary system. The predicted pressure loss for the thermal bus, assuming a worst-case heat load of 3 kW, is less than 0.3 psi, providing pumping margins greater than 40%.

System ORU radiator capacity was analyzed for both POP and COP missions. In general, ORU's are mounted on the $(-Z)$ zenith side of the carrier. Because the $(-Y)$ and $(-Z)$ sinks are cold space, ORU radiators with $-Y/-Z$ FOV's exhibit a relatively constant 400-W reject capacity (with a 20°C radiator) through the required sun-angle variation. Heat rejection for ORU's having a less desirable $+Y/-Z$ FOV is limited to 250 W because of sun incident on the $+Y$ radiator during $\beta = 10$ deg (1330 nodal crossing) operation, where β is the angle between the sun and the orbital plane.

Co-orbiting telescope missions have strict sun-avoidance criteria to minimize heating of actively cooled telescope apertures. ORU's mounted on the "lower" hemisphere of the telescope will not have a solar view and provide heat rejection in excess of 400 W. In contrast, ORU's mounted on the "upper" telescope hemisphere are subjected to solar radiation and suffer a reduction in heat-rejection capacity to approximately 160 W.

Summary

Mission requirements for the Space Station free-flying platforms have shaped the TCS configuration. Modularity, heat acquisition/rejection, longevity, and temperature tolerances have led, through a systematic trade study matrix, to a hybrid TCS that employs a central core fluid loop for thermal control mechanisms for the platform subsystems. Platform orbital orientation has been favorably exploited for radiator and experiment cooler location. The platform dimensions and heat-transport requirements are amenable to a capillary pumped fluid loop, with ammonia as a working fluid.

Analysis of the total platform TCS indicates that, for the worst-case combination of environmental, equipment operation, and material property assumptions, the thermal design requirements are accommodated with comfortable margin.

References

[1]Braun, C. E., Hartshorn, K., and Pergament, S., "Thermal Design of the Space Station Force-Flying Platforms," AIAA Paper 88-2698, June 1988.

[2]*Trasys II User's Manual*, Martin Marietta, MCR-73-105 (Rev. 5), June 1983.

SP-100 Space Reactor Power System

A. Kirpich,* G. Kruger,† D. Matteo,‡ and J. Stephen§
General Electric Company, Princeton, New Jersey

Introduction

This chapter describes the generic flight system (GFS) design for a 100-kWe space reactor power system (SRPS) undertaken by the General Electric Astro-Space Division as part of the SP-100 Ground Engineering System (GES) Program under contract with the Department of Energy. Key objectives of the GES Program are to identify the principal technologies associated with space nuclear power, establish development programs where technology shortfalls exist, and validate that the needed technologies are ready for flight implementation.

The SP-100 Program as a whole has the objective of establishing, as a national resource, the capability of producing space power in the range of 10–1000 kWe utilizing nuclear reactor technologies in conjunction with appropriate methods of energy conversion. Studies conducted over the past 5 yrs have converged on several key selections that have provided the basic framework against which the GFS design is evolving. Principal among these are 1) the selection of a lithium liquid-metal-cooled reactor constructed of refractory metals permitting operation in the range of 1300–1400 K; 2) heat transport by lithium circulation using thermoelectrically driven liquid-metal pumps; 3) thermoelectric power conversion; and 4) waste heat rejection at approximately 800 K through lithium circulation to potassium heat pipe radiators.

Copyright © 1989 by the American Institute of Aeronautics and Astronautics Inc. No copyright is asserted in the United States under Title 17, U.S. Code. The U.S. Government has a royalty-free license to exercise all rights under the copyright claimed therein for Governmental purposes. All other rights are reserved by the copyright owner.
*System Design Manager, SP-100 Program, Astro-Space Division.
†Nuclear Integration Manager, SP-100 Program, Astro-Space Division.
‡Space Subsystems Manager, SP-100 Program, Astro-Space Division.
§Nuclear System Project Engineer, SP-100 Program, Astro-Space Division.

The GFS consists of a highly integrated set of thermal-hydraulic subsystems. To maintain design consistency among these subsystems, interfaces must be clearly delineated and their characteristics defined. Considering the high-temperature conditions of operation and the transport phenomena associated with coolant circulation, numerous thermal-hydraulic considerations enter into the GFS definition. Various thermal-hydraulic analytical procedures have been utilized in establishing the design for the reactor, ducting, hot-side and cold-side heat exchangers for the thermoelectric power converter, circulating pumps, and heat pipe radiators. Considerable attention has been given to the effect that micrometeoroid protection has on optimizing the thermal-hydraulic design. Additionally, material compatibility is a major consideration that must be carefully monitored throughout the design process.

The GFS design described in this section represents the current definition for solving many of the identified thermal-hydraulic concerns. Some of the more unique thermal-hydraulic concerns addressed thus far in the program are described briefly below.

Startup Thaw

For safety reasons, the SRPS is launched with the lithium circulant in a frozen condition. Thaw must be accomplished after the SRPS and its attached payload are injected into the operational orbit. The thaw process involves heat generation within the reactor and subsequent heat delivery to remotely located piping and components. Results to date, discussed more comprehensively in Chapter 3, suggest that complex thermal and physical interactions must be dealt with. Such complexities as volumetric phase change and prethaw void formation and placement must be carefully considered. The definition of an appropriate thaw scheme is clearly fundamental to the entire thermal-hydraulic concept.

Heat Transport

The pumping of lithium to deliver reactor thermal power involves the use of thermoelectric-electromagnetic (TEM) pumps. Such pumps have been previously designed but are uniquely applied to the SP-100 GFS in that they are designed to circulate lithium for both primary reactor heat delivery and secondary waste heat removal within a single pump body.

Heat Rejection

Potassium heat pipes arranged into radiator panels provide the means for waste heat removal. The heat pipe evaporator sections are bonded to ducts through which lithium is circulated for delivering the waste heat. Special considerations pertain to maximizing heat pipe reliability, providing protection against micrometeoroids and other space debris, and designing the duct to be easily thawed during the startup sequence.

Gas Separation

A unique issue concerns the generation of helium within the primary lithium loop resulting from nuclear reactions within the reactor vessel. Methods of helium separation are necessary to ensure that heat-transfer rates are not degraded and that pumping capability is unimpaired.

The following section provides an overview description of the SP-100 GFS.

Overview Description

The deployed, or operational, configuration of the 100-kWe GFS is shown in Fig. 1. The overall length from the reactor tip to the user interface plane is 23.2 m. The radiation shadow shield just aft of the reactor attenuates nuclear radiation within a 17-deg half-cone angle. All equipment of the SRPS is contained within this cone in order to achieve acceptable radiation levels at the user interface plane. The power generation module (PGM) consists of the forward main body, which houses the reactor, shield, pumps, power converters, heat-rejection radiators, and other miscellaneous equipment. The user interface module (UIM) is comprised of the separation boom, the shunt dissipator, and the equipment module, which is a housing in which power conditioning equipment, batteries, and system controls are mounted. The separation distance reduces both the radiation

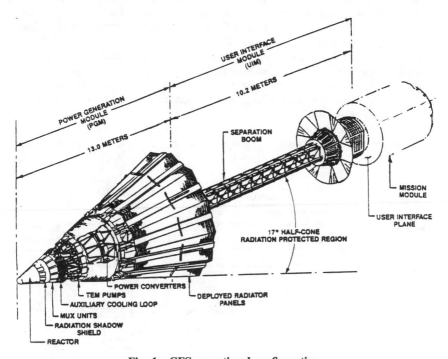

Fig. 1 GFS operational configuration.

and thermal environments to which the mission module and UIM electronic equipment are subjected.

A simplified hydraulic schematic of one of 12 similar cooling segments is shown in Fig. 2. Lithium is used as the fluid because of its favorable mass, specific heat, and vapor pressure properties. Its frozen condition during launch provides safety advantages. After insertion into the required orbit, the lithium distributed through the system is thawed by a heat pipe network that derives its heat from the reactor. Upon full thaw the primary loop high-temperature lithium is circulated through the reactor (which serves all 12 segments) and is routed to the inlet of a TEM pump. The pump delivers the fluid to the hot side of a thermoelectric power converter and then returns the fluid to the reactor. Heat given up by the fluid passes through thermoelectric elements in the power converter and TEM pump. In the process, some of the heat is transformed to electricity, and the remaining heat is transferred to a secondary low-temperature lithium loop for transport to a heat pipe radiator and rejection to space. Electricity developed within the power converter is combined with that from the outer segments and delivered as the system output (100-kWe rating) at 200 V. A shunt regulator maintains voltage within regulation limits by absorbing excess power not demanded by the user load. Electricity developed within the TEM pump operates in conjunction with a self-induced magnetic field to produce the hydraulic pumping force. The TEM pump is dual-acting by virtue of producing hydraulic pumping for both the primary and secondary loops within a single pump body. All 12 pumps (1 per segment) operate from a single fluid space serving the primary loops while their secondary loops are totally independent. Because of their higher vulnerability, the primary loops are configured to be compact to minimize exposed area and micrometeoroid protection requirements. Secondary loop vulnerability is less severe, since the loss of a loop is essentially limited to the contribution of the associated segment.

Table 1 enumerates the principal features of the system concept in terms of the nuclear and space subsystems that comprise the overall GFS.

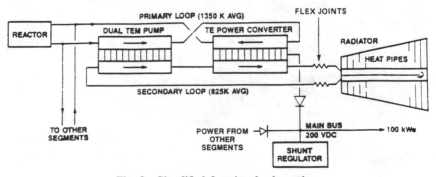

Fig. 2 Simplified functional schematic.

Table 1 GFS subsystem features

Nuclear Subsystems		Space Subsystems	
Reactor		*Power Conversion*	
Neutronics	Fast spectrum	Thermoelectrics	SiGe/GaP
Fuel	UN	Thermal coupling	
Coolant	Lithium	—hot-side	Conductive compliant pad
Vessel and pin clad	PWC-11	—cold-side	Conductive compliant pad
Control	12 Radial reflectors	High voltage insulator	Single crystal alumina
Safety shutdown	7 In-core rods		
Re-entry shield	Carbon-carbon	*Heat rejection*	
		Method	Pumped lithium loops (12)
Shield		Circulation	Dual TEM pumps (12)
Neutron attenuation	Lithium hydride	Radiator	Potassium heat pipes
Gamma-ray attenuation	Tungsten		
		PCC&D	
Heat transport		High power bus	200 VDC
Method	Pumped lithium loop (12)	Low power bus	28 VDC
Circulation	Dual TEM pumps (12)	Regulation	Full shunt
Gas separation	Surface tension barrier	Energy storage	Primary/secondary batteries
		Critical loads bus	Triply redundant supply to RI&C
Instrumentation & control			
Sensors		*Structure/mechanisms*	
Drives		Extension boom	
Electronics/mux units		State-of-art methods and devices	
Multiple redundancy			
		*Auxiliary Cooling Loop**	
		In-core bayonet HX's	
		Dual TEM pump	
		Power converter	
		Radiator	
		*Elements distributed among S/S's	

The reactor subsystem employs a UN-fueled, fast-spectrum, lithium-cooled reactor made of PWC-11 niobium alloy. Control is exerted by 12 hinged radial reflectors that surround the reactor vessel. Seven in-core safety rods are used, all of which must be extracted to permit startup. A carbon-carbon heat shield surrounds the reactor and ensures that it will remain intact upon atmospheric re-entry.

The shield subsystem provides a reduced radiation shadow with a 17-deg half-cone angle with sufficient attenuation to provide a lifetime neutron fluence of 10^{13} neutrons/cm^2 and a gamma dose of 5×10^5 rads (Si). Encased lithium hydride serves as the principal neutron shield and tungsten layers as the gamma shield.

The primary heat-transport subsystem (PHTS) distributes the reactor lithium coolant by means of 12 dual-action TEM pumps. Each TEM pump consists of hot-side and cold-side ducts with thermoelectric (TE) elements sandwiched in between and a self-inducing magnetic structure that, in conjunction with the TE-developed current, produces hydraulic pumping in both primary and secondary loops. The PHTS includes the primary loop piping and six gas separator/accumulator (GSA) units, one each for two TEM pumps. The GSA units separate helium bubbles generated within the reactor by nuclear processes and prevent their circulation through the primary loop. Pressure from an initial helium charge and the later pressure buildup due to generated helium serve to provide an accumulator function for accommodating lithium fluid expansion. The PHTS also includes a thaw heat pipe (THP) system for initially thawing the lithium during system startup. Twelve reactor THP's derive heat from the reactor vessel wall and deliver heat to additional branched THP's. The use of conductive saddles and straps along with the strategic placement of insulation results in the proper transfer of thaw heat.

The reactor instrumentation and control system (RI&C) consists of the sensors and drivers for exerting control over the reactor. Temperature, flow, and reflector position are the principal sensed parameters. Motors and drive mechanisms are the means for changing reflector settings. Separate drives are used to extract safety rods. Signal conditioning and multiplexing is accomplished by multiplexer/demultiplexer units located behind the reactor radiation shield. These units both improve the signal-to-noise ratio of transmitted signals and reduces the cabling for delivering the signals to the reactor controller located in the user interface module.

The power conversion subsystem (PCS) uses conductively coupled SiGe/GaP thermoelectrics (TE's) as the basic conversion method. The TE's consist of N—and P—legs which, when appropriately interconnected, produce a rated output in excess of 105 kWe at 208 V. The TE's are assembled into cells with provisions for high heat conduction, electrical insulation, and expansion compliancy. They are then arranged into planar layers and sandwiched between cold-side and hot-side heat exchangers. With appropriate manifolds the heat exchangers are plumbed into the primary and secondary lithium loops. A high-temperature material has been developed as the high-voltage insulator material; it must limit the total leakage current to an acceptably low value at a standoff voltage in excess of 100 V at temperatures up to 1350 K.

The heat-rejection subsystem (HRS) removes waste heat from the power converters by means of the 12 independent secondary loops. Pumped circulation is provided by the dual-action TEM pumps described under the PHTS. The waste heat is transported to a longitudinal radiator duct to which transverse heat pipes are brazed. The heat pipes consist of coextruded beryllium-titanium, D-shaped tubes with internal wicking and potassium working fluid. The waste heat transport duct has supply and return passages with a unique fluid interchange arrangement for accomplishing thaw during system startup.

The power conditioning control and distribution subsystem (PCCD) regulates the voltage of the main output by means of a shunt regulator. The PCCD also delivers 300 W of 28-V power to the user as well as providing 28-V power for RFS housekeeping needs. A complement of primary and secondary batteries serves startup and restart needs. Extensive redundancy is applied throughout the subsystem assuring uninterrupted operation in the face of single-point failures.

The mechanical/structural subsystem encompasses all structural support members, mechanisms, and thermal control needed to meet launch and operational load requirements. The principal elements of this subsystem include a structural framework for supporting the reactor, shield, TEM pumps, power converters, and radiator panels; the UIM extension boom; and the equipment module assembly for housing PCCD and RI&C electronic equipment. The subsystem also includes mechanisms for radiator panel deployment and launch lock releases that permit free expansion of piping after completion of the launch phase.

The auxiliary cooling loop (ACL) serves to maintain heat removal from the reactor following a loss-of-coolant accident (LOCA). The ACL consists of an independent cooling loop with in-core bayonet-type heat exchangers replacing some of the fuel pins in the reactor. With loss of primary loop coolant, the ACL will continue to function and remove sufficient heat to prevent excessive fuel temperatures. The ACL is treated as a functional entity whose elements are distributed among the eight designated nuclear and space subsystems.

Performance

GFS physical and performance characteristics are summarized in Table 2. The net delivered power at the user interface is 100 kWe at 200 V and 300 We at 28 V.

Power margin estimates as a function of mission time are shown in Fig. 3. The upper line pertains to having 12 operative segments, and the lower line pertains to having 11 operative segments. Optimum mass considerations favor the strategy of permitting the loss of 1 out of 12 segments rather than using added protection to ensure that all segments remain intact within a specified probability. The ability of the 11 segment case to furnish rated power at end of mission (10 yr) reflects the use of this strategy.

Table 2   GFS performance summary

Physical characteristics		Performance characteristics	
Mass (kg)	3600 to 5400[a]	Delivered power	100 kWe @ 200 V 300 We @ 28 V
Radiator area, one-side (m²)	106.4	Thermal power (MWt)	2.4
Power converter area (m²)	6.6		
Stowed length (m)	6.85	Reactor outlet temperature (K)	1375[b]
Deployed length (m)	23.1	Radiator temperature, av (K)	800[b]

[a]Depends on space environment and mission life factors.
[b]End-of-mission conditions.

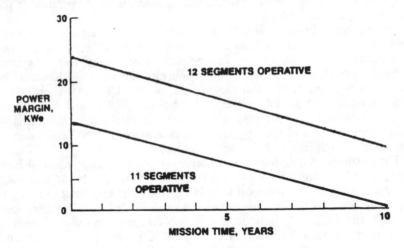

Fig. 3   Power performance margin.

User Interfaces

Interface characteristics of the GFS, from a user standpoint, are summarized in Table 3. The characteristics are provided with respect to a hypothetical user plane that is perpendicular to the GFS axis and is located 22.5 m behind the centerline of the reactor core. The user plane diameter is 4.5 m. Figure 4 shows the GFS in the stowed configuration as it would be mounted in the NSTS (Shuttle) cargo bay. The GFS is cantilever mounted to the mission module (payload) by means of twelve 19-mm separation bolts equally spaced on a 3.35-m bolt circle. The GFS user interface module is hard-mounted to the mission module at the user interface plane. The user electrical interfaces are located in the aft section of the UIM as shown in

Table 3 Generic flight system external interfaces

Interface parameter	Characteristic
User plane location	22.5 m behind reactor core centerline
User plane diam	4.5 m diam
Mission module mechanical interface	Hard mount to user interface module
Resonant frequency of boom-mounted PGM	$\geqslant 1$ Hz
Thermal flux at user plane	0.14 W/cm^2
Radiation at user plane	10^{13} Neutrons/cm^2; 5×10^5 rads (Si)
Load following	Rapid response full shunt
Main bus	100 kWe, 200 V $\pm$ 5%
Secondary bus	300 We, 28 $\pm$ 7 V
Telemetry	16 kbps
Command	2 kbps
Launch	National Space Transportation System (Shuttle)

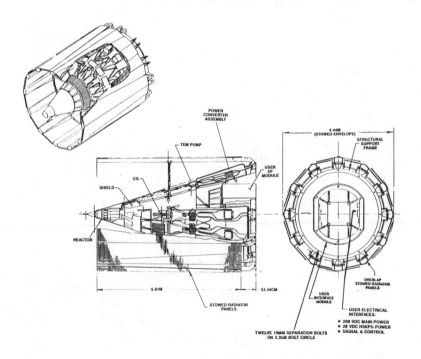

Fig. 4 GFS stowed configuration.

the figure. Boom extension first requires that the mission module be separated from the 3.35-m ring that is part of the power generation module structural support frame. This is accomplished by the 12 separation bolts. This is followed by boom extension in which the UIM/mission module assembly is separated from the PGM. In the fully deployed position the resonant frequency of the boom-mounted PGM is greater than 1 Hz.

Safety

The RFS includes a number of key design features for ensuring safe operation during all mission phases:

The reactor is not operated until the SRPS/mission module has been inserted into its proper operational orbit. Thus, launch accidents or other abort situations will not involve the dispersion or disposition of fissioned material.

Launch is accomplished with the lithium frozen, avoiding the hazards of a piping or vessel rupture.

The reactor employs 2 independent methods for shutdown during operation by the release of any 4 of 12 control reflectors or by the insertion of any 2 of 7 safety rods.

The reactor includes an auxiliary cooling loop for ensuring adequate core heat removal in the event of a loss-of-coolant accident.

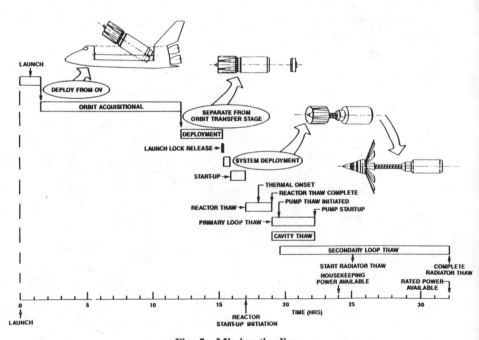

Fig. 5   Mission timeline.

The reactor is protected by a heat shield to ensure that it remains intact in the event of accidental atmospheric re-entry.

Analysis indicates that the reactor will remain subcritical in mission aborts including launch vehicle explosions and following impact and burial or full water submersion.

Operation

The GFS is operationally compatible with the requirements for launch and ascent, startup, normal mission operation, operational shutdown and restart, end-of-mission shutdown, and final shutdown.

Figure 5 illustrates the launch, deployment, and startup sequence. Following the boost, orbit acquisition, and booster separation phases, the SRPS with its attached payload will operate autonomously on internal batteries. After boom extension, radiator panel deployment, and the release of launch locks, reactor startup is initiated about 17 h after launch. The reactor is fully thawed about 2 h later and is raised to a bulk temperature of 950 K. At this temperature, thaw heat pipes attached to the reactor vessel light off and transport heat to the primary loop, TEM pumps, and power converters. Five hours after startup the TEM pumps are fully thawed and receive sufficient heat to initiate pumping. The radiator is fully thawed 15 h after startup initiation. Electrical power is initially generated about 7 h after startup, relieving the demand on the internal batteries. Power buildup occurs gradually until full rated output is generated 15 h after startup.

Advanced Multimegawatt Space Nuclear Power Concepts

John A. Dearien* and Judson F. Whitbeck*
Idaho National Engineering Laboratory, Idaho Falls, Idaho

In response to the need of the Strategic Defense Initiative (SDI) and long-range space exploration and extraterrestrial basing by NASA, concepts for nuclear power systems in the multimegawatt levels are being designed and evaluated. The requirements for these power systems are being driven primarily by the need to minimize weight and maximize safety and reliability. This chapter will discuss the present requirements for space-based advanced power systems, technological issues associated with the development of these advanced nuclear power systems, and some of the concepts proposed for generating large amounts of power in space.

Design Requirements for Advanced Power Systems
There are presently several areas that reflect the needs of different groups and drive the design requirements of space-based power systems. A brief summary of these groups and their requirements is given below.

Strategic Defense Initiative
Advanced weapon systems for the SDI require power in the multimegawatt range with the unique requirement that the weapon, and therefore the power system, may be required to remain in orbit for years, unused, and then on a very short notice, be called on to operate for hundreds of seconds at full power. This "burst" mode of operation results in a much different set of design requirements than a normal baseload power station.

---

*Senior Program Specialist, Multimegawatt Project, Technical Support Office.

Most of the weapon concepts for the SDI system utilize liquid hydrogen to cool the weapon, and this hydrogen can be used, after cooling the weapon, as a coolant for the power system. In the case of nuclear power sources, the availability of this onboard hydrogen, in a sense "free" after it has been used to cool the weapon, drives a power system designer quite naturally to the use of an open-cycle Brayton system. A nuclear reactor heats the hydrogen, drives a turbine/generator with the resulting hot gas, and exhausts the gas overboard. Several of the concepts described in this section are of this type.

Many of the weapon concepts of SDI have very sophisticated sensors and control systems onboard, and there is concern about the amount and composition of exhaust gases that can be tolerated in the vicinity of the weapon and not affect operation of the weapon system. For this reason, there are concepts being addressed that can deliver multimegawatt levels of power and not produce effluents. Concepts of this type are also discussed in this section.

SDI has the generic requirements, as all potential users do, of safety, lightest possible weight because of high launch costs, and extremly high reliability, both in the dormant stage and the high-power burst mode.

NASA

NASA is continuing an extensive planning effort that is directed toward establishing and maintaining permanent human presence beyond the confines of the Earth's gravitational field. The missions within this planning effort include expeditions to the Mars system, science outposts on the lunar surface, evolutionary bases on the lunar and Mars surfaces, and expeditions to near-Earth asteroids. The potential leverage inherent in reduced mass associated with nuclear power and propulsion systems is being studied to provide a basis for directing appropriate nuclear technology research and development.

The potential applications for nuclear technology can in general be classified in two categories; open-cycle, short-duration bursts for propulsion, and closed-cycle, steady-state, long-term production of thermal or electrical power for surface support or electric propulsion. Specific applications could include

1) direct thermal propulsion for manned and cargo vehicles to and from the Mars system;

2) electric propulsion for lunar cargo vehicles and Mars system cargo vehicles;

3) surface base electrical power for environmental and logistics support;

4) surface base electrical or thermal power for in situ resource processing; and

5) orbital electrical power generation for subsequent beam transmission to space-based or surface-based loads.

Generic requirements for power and propulsion systems of pertinence to NASA are consistent with those for SDI applications, in particular with respect to safety, reliability, and mass. The presence of humans in proximity to nuclear power and propulsion systems gives additional emphasis to safety considerations in the design, deployment, operation, and disposal of such systems for NASA applications.

Technology Issues

Most of the technology issues being addressed are a result of the need to develop high-power systems with the minimum possible weight. With launch capacity limited to less than 30,000 kg at present in the Shuttle, or a future capacity of 50,000–60,000 kg with the Advanced Launch System (ALS), and costs that are measured in thousands of dollars per kilogram, there is a tremendous motivation to make these systems as light as possible. This drive toward lightweight systems is reflected in the use of high operating temperatures to increase cycle efficiency, the use of high-strength, high-temperature metals and composites, and the development of new and innovative concepts in the areas of heat rejection and power conversion/power conditioning.

Safety is the paramount concern of all concepts being designed for power in space, and this concern is reflected in all aspects of the design, development, testing, launch, usage, and ultimate disposal of any proposed system. This concern has particular effects in the thermal hydraulic design of a system. Many of the proposed closed-cycle systems use liquid metals as a reactor coolant. Safety in launch requires that these systems be launched frozen and then melted in space. This requirement puts a formidable design task on any system, not only from having to perform the initial startup after launch, but also from having to deal with the potential for an operational shutdown, freezeup, and restart while in orbit.

Heat rejection for all systems, open or closed, is a primary issue due to the constraints of weight and microgravity. The thermal hydraulic aspects of the systems must address the cooling of not only the reactor but the power conversion and power conditioning equipment as well. This does not seem to be a big problem at this time for the SDI open-cycle systems because of the availibility of excess liquid hydrogen from the weapon cooling system. The design issues mainly concern delivering the appropriate amount of hydrogen to the particular component and then disposing of it in a manner that will not interfere with operation of the weapon or other sensitive equipment on the platform. For systems utilizing superconducting generators, an additional thermal hydraulic task of dealing with a separate liquid helium system is required.

Heat rejection for closed-cycle systems is a far more difficult task, since all heat rejection must ultimately be made by radiation to space, and the present heat-transfer equipment (heat exchangers, radiators, recuperators, etc.) are heavy and voluminous.

Technology issues in space generally result from spacecraft (or base) requirements, microgravity, safety considerations, and in some cases the use of new components or a component's use outside the normal experience range. The problems differ greatly depending on whether an open or closed cycle is being considered. Closed Rankine cycles, which utilize phase change fluids, provide a variety of thermal-hydraulic concerns that do not exist with Brayton cycles, either open or closed. Some of the major impacts of these items on closed- and open-cycle systems are highlighted below.

Microgravity Considerations

1) The use of a phase change fluid in microgravity requires a means to manage the two phases. Thus separators are a part of the system.

2) Phase change power systems have condensers that must operate where shear forces control condensation since no falling film condensation can exist.

3) Obtaining adequate net positive suction head (NPSH) for the main feed pump in a closed Rankine cycle requires special techniques, i.e., rotary fluid management device, jet pumps.

4) Hydrogen acquisition, NPSH development, and pumping are major concerns for open-cycle systems since generally no acceleration will exist when the system is started.

5) Surface tension controlled circulation flows may exist in some closed thermal energy storage systems.

Spacecraft-Related

1) The SDI missions may require that a spacecraft accelerate during operation of the power system. This may have the effect of stalling or reducing flows in a component where the driving head is small, i.e., heat pipe, or reactor loop with very-low-pressure drop. Thus, maneuvers may cause flow and thermal transients on the system and must be considered in the design of components.

2) In open-cycle SDI power systems the weapon systems (primarily accelerator and RF cooling paths) act as preheaters with respect to the power system. Thus, the thermal-hydraulic integration of weapon cooling with the power system is a major factor in the thermal design of the power system. This is especially true for the normal rapid startups and shutdowns and thermal implications of load rejections resulting from faults that will impact the state of the hydrogen entering the power system.

3) SDI has very restrictive vibration requirements that dictate the fluid conditions in some portions of the weapon system. Supercritical hydrogen is required for accelerator cooling to eliminate the flow-induced vibration that could result from boiling in the coolant passages and thus partially establishes the reactor system fluid inlet conditions.

4) The need for an SDI system to be tested regularly and to tolerate very rapid load rejection has prompted the use of dummy electrical loads that may be hydrogen cooled in open-cycle systems. These dummy loads are heaters that are integrated into the hydrogen flow path and thus provide an additional thermal consideration.

5) The concerns of hydrogen acquisition and pumping noted earlier are increased because of SDI spacecraft maneuvering requirements. Under these conditions the potential for flow reduction, instability, and interruption to both the cooling circuits and the power system could occur as a result of ineffective fluid acquisition within the tanks and/or hydrogen

pump cavitation. Any pumping system within the power system, as with a closed cycle, must address this problem. Thus, the microgravity considerations are often exacerbated.

Safety-Related

1) Launch safety considerations require that liquid metal systems be frozen during launch. This requirement leads to complex thaw systems and analytical methods to evaluate the process.

2) The multimegawatt space reactor power system will be designed to ensure that the reactor remains structurally sound for all normal operating modes and credible accidents, including loss of primary flow or coolant.

3) The multimegawatt space reactor will be designed to remain subcritical even if a launch accident or inadvertent re-entry were to occur. It is also the objective of the space reactor program to avoid, or if that is not possible, minimize, the use of toxic materials, such as beryllium, which could be dispersed during a launch accident or an inadvertent re-entry.

4) Multimegawatt space reactor power systems will not be operated until planned orbits are achieved. The lifetime of an orbit is usually defined as the time it takes a spacecraft in a specific orbit to re-enter the Earth's atmosphere and is a direct function of orbit altitude. The planned orbit lifetime (and altitude) of a space multimegawatt reactor system will be carefully selected, either initially or by subsequent boost, to ensure sufficient time for radioactive materials to naturally decay in space to acceptable levels. The selected orbit altitude (lifetime) will depend on the operating history and fission product inventory of the reactor at the end of life.

5) The functional and operating requirements for multimegawatt space reactors specifies that the reactor, either by placement in an initial long-life orbit or boost to high orbit after operation, not re-enter the Earth's atmosphere. However, as an added safety measure, design specifications for the reactor also require that the reactor re-enter intact in the event of an inadvertent re-entry. This will result in reactor burial in soil, pavement, or water upon impact and will ensure that reactor material would be contained within a small area.

6) Concerning environment for manned systems, reactor systems developed to date are for unmanned power sources, and the only radiation shielding has been for protection of electronic equipment. If operational missions involving manned vehicles advance, shielding weights will also likely increase to provide additional protection for humans.

Advanced Multimegawatt Space Nuclear Power Concepts

A number of reactor concepts have been proposed as part of the Multimegawatt Program (MMW) jointly sponsored by the U.S. Department of Energy and the Strategic Defense Initiative Office (SDIO). The requirements driving the design of these concepts are primarily those of the SDI, but there are several aspects of the concepts that are consistent with

Table 1 Proposed reactor systems

Company	Power	System
Boeing	tens of MWe	Open-cycle Brayton
General Electric	tens of MWe	Open-cycle Brayton
Westinghouse	tens of MWe	Open-cycle Brayton
Grumman	hundreds of MWe	Open-cycle Brayton
General Atomics	hundreds of MWe	Closed-cycle thermionic reactor with fuel cells
Rockwell	hundreds of MWe	Closed-cycle liquid metal Rankine with batteries

the requirements of NASA. Table 1 is a listing of these concepts with a description of the type of system, power range (exact power levels are classified), and the developing vendor. The concepts in this table are described in detail in the following sections.

Boeing

System Description

Boeing has developed a hydrogen-cooled open Brayton cycle system in the tens of MWe range. This is a new reactor and uses a fuel-pin-type core. The system utilizes twin counter-rotating turbines and alternators and a balanced thrust exhaust system. The system layout is shown in Fig. 1.

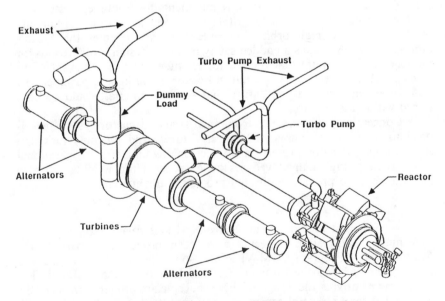

Fig. 1 Boeing open Brayton cycle system component layout.

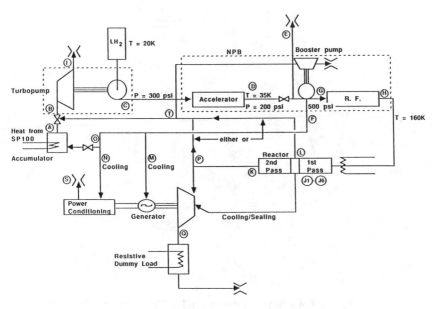

Fig. 2   Boeing open Brayton cycle system flow diagram.

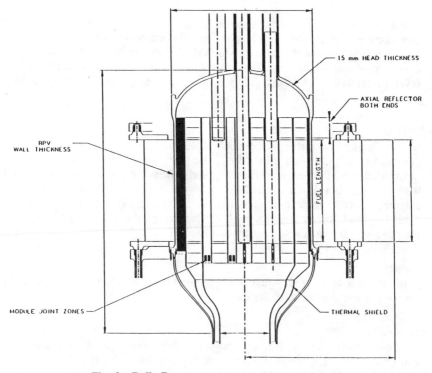

Fig. 3   Rolls Royce two-pass reactor arrangement.

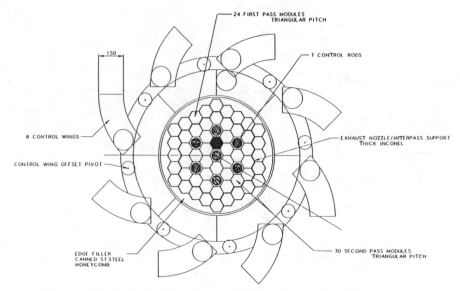

Fig. 4   Rolls Royce two-pass reactor core cross section.

Figure 2 shows the system schematic and again illustrates the configuration common to the open-cycle systems. The schematic notes the two-pass flow configuration of the core in which the hydrogen enters the reactor and flows up through the outer annulus of fuel pins to the upper plenum, reverses direction, and then flows down through the center array of fuel pins.

Reactor

The Boeing concept uses a British Rolls Royce designed reactor, hydrogen-cooled with a fuel pin core. The fuel pins are contained within hexagonal modules on a triangular pitch. Reactor control is maintained with movable beryllium control wings on the outside of the pressure vessel. Figure 3 shows the reactor, and Fig. 4 shows a cross section through the core. Figure 3 shows the flow path of the two-pass core. Design operating temperature at the reactor outlet is 1200 K.

Turbine

Boeing chose an eight-stage turbine operating at 15,000 rpm. Figure 5 shows the basic layout of the turbine.

Generator

The generator for the Boeing design is a brushless wound rotor machine chosen for its reliability and operational efficiency. Figure 6 shows the overall features of the generator.

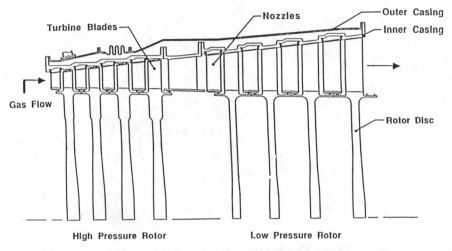

Fig. 5   Boeing turbine general arrangement.

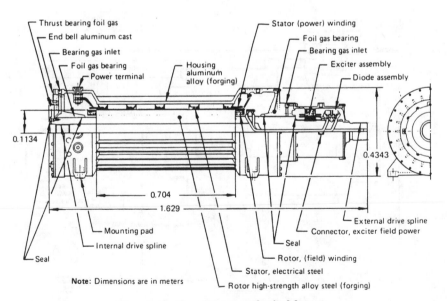

Fig. 6   Boeing alternator physical layout.

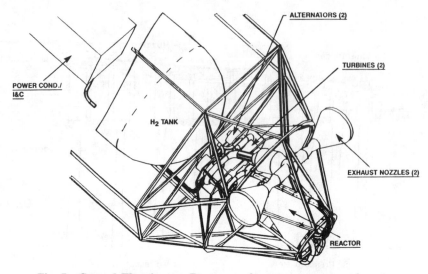

Fig. 7　General Electric open Brayton cycle system component layout.

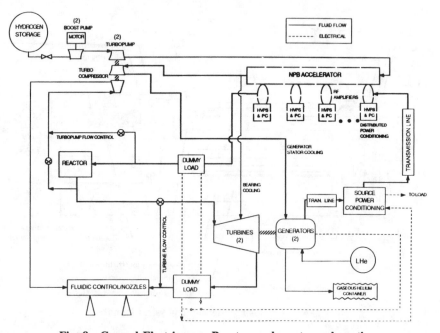

Fig. 8　General Electric open Brayton cycle system schematic.

General Electric

System Description

General Electric has designed a reactor system in the tens of megawatts range using a derivative of the reactor (the 710 reactor) designed for the PLUTO program in the 1960's. This reactor is a hydrogen-cooled open Brayton cycle, and the system configuration is shown in Fig. 7.

Figure 8 shows the system schematic. The flow diagram is typical of all of the open-cycle systems with the addition of the liquid helium system for cooling of the superconducting generator.

Reactor

GE uses a derivative of the 710 reactor originally developed as part of the PLUTO program in the 1960's. The fuel for this reactor is a ceramic-metal (cermet) mixture of UO_2 and Mo and sheathed with an MoRe clad. Figure 9 shows the reactor, and Fig. 10 provides details of the fuel configuration and clad. The design operating temperature for this reactor is 800 K. Testing of fuel elements for the 710 program produced data on this fuel type at temperatures of up to 2800 K, in pile tests of up to 10,000 h, and hundreds of thermal cycle tests.

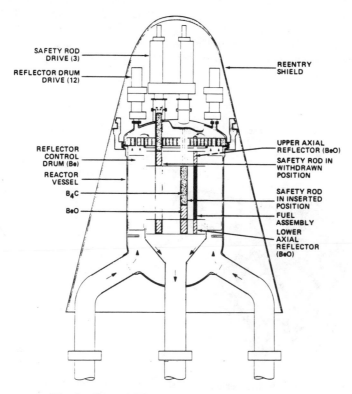

Fig. 9   General Electric reactor core assembly.

Turbine/Generator

GE uses two seven-stage counter-rotating turbine/generators integrated into a single mechanical unit and turning at 12,000 rpm. The generator is superconducting and is cooled by an onboard supply of liquid helium. Figure 11 shows the integrated turbine/generator configuration.

Westinghouse

System Description

Westinghouse has developed a system designed to provide electrical power in the range of tens of megawatts using a NERVA derivative reactor operating with hydrogen in an open Brayton cycle. Figure 12 shows the concept and illustrates the location of major system components within a protective thermal and space debris shield. The power system is designed such that it can be separated from the user platform. This enables the power system to escape a distressed platform, allows for separate disposal, and enables on-orbit replacement of a faulty power system.

Figure 13 shows the system schematic and illustrates several of the features that are common with the open-cycle systems for this SDI application. Hydrogen for reactor cooling is drawn from a large hydrogen tank at cryogenic temperature, pumped through the weapon to cool it, and then utilized in cooling of the reactor and power generation. Counter-rotating turbines and generators and balanced exhaust nozzles are employed to keep power system mechanical loads on the weapon platform to a minimum.

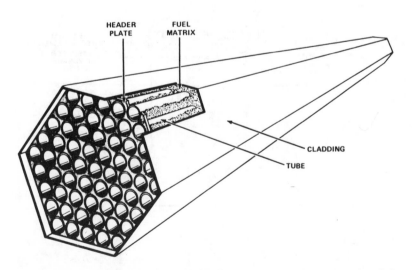

Fig. 10   General Electric reactor fuel element with 61 coolant channels, clad and matrix details.

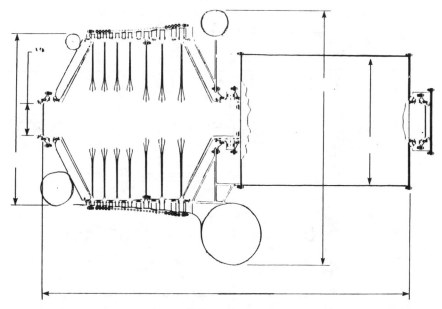

Fig. 11   General Electric power turbine design.

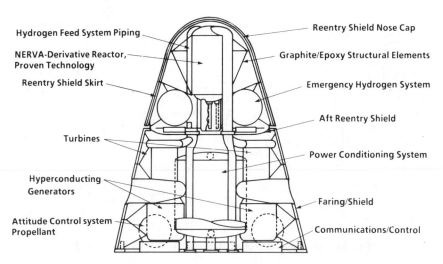

Hydrogen Feed System Piping

NERVA-Derivative Reactor, Proven Technology

Reentry Shield Skirt

Turbines

Hyperconducting Generators

Attitude Control system Propellant

Reentry Shield Nose Cap

Graphite/Epoxy Structural Elements

Emergency Hydrogen System

Aft Reentry Shield

Power Conditioning System

Faring/Shield

Communications/Control

Fig. 12   Westinghouse open Brayton cycle system component layout.

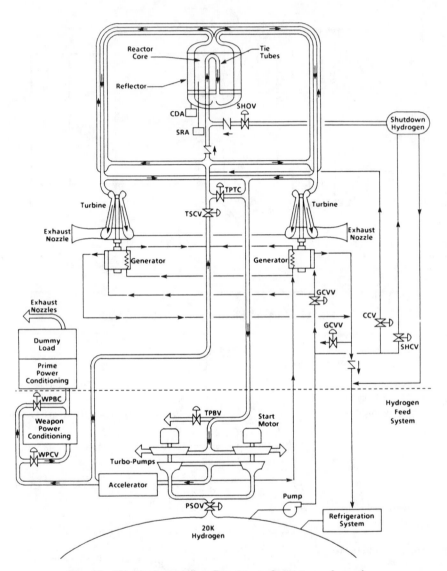

Fig. 13   Westinghouse open Brayton cycle system schematic.

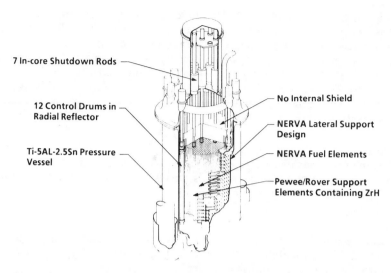

Fig. 14   Westinghouse ZrH and graphite moderated reactor configuration.

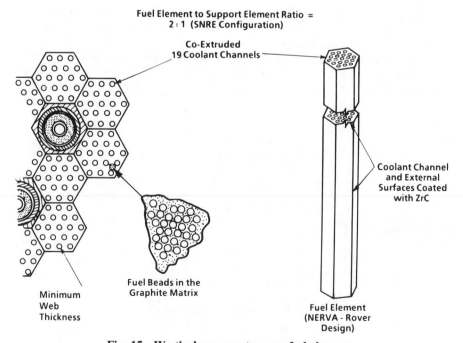

Fig. 15   Westinghouse reactor core fuel element.

Reactor

This concept uses a NERVA derivative reactor as a power source. The core is comprised of a large number of hexagonal graphite rods containing encapsulated fuel pellets of enriched uranium. Each graphite rod has 19 small flow channels through which the hydrogen coolant flows and is heated. Figure 14 shows the reactor configuration, and Fig. 15 shows the detail of the core elements. The design operating temperature of this reactor is 1150 K. This basic design has substantial operational data at higher power levels and higher temperatures as a result of the NERVA program.

Turbine

Westinghouse has chosen an eight-stage turbine and optimized it for use with hydrogen. Figure 16 shows this turbine and illustrates the details associated with such a machine.

Generator

Westinghouse has chosen a hyperconducting generator as a candidate generator. The availability of liquid hydrogen makes hyperconducting generators a logical choice for consideration.

Grumman

System Description

Grumman developed a hydrogen-cooled, open Brayton cycle system to deliver electrical power in the range of hundreds of megawatts. The concept utilizes a particle bed reactor in which the fuel is in the form of very small (500 μ) particles in which enriched uranium is encapsulated within protective coatings. The basic fuel particles have been used in gas reactors for a number of years. Use of the particles in this particular reactor configuration

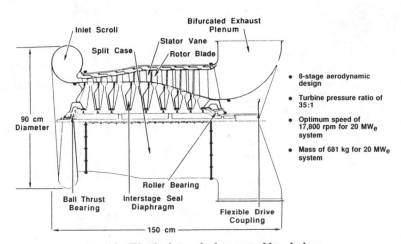

Fig. 16   **Westinghouse hydrogen turbine design.**

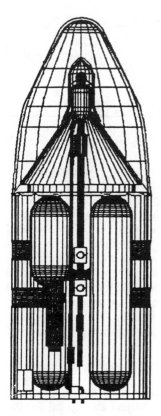

Fig. 17   Grumman open Brayton cycle system concept configuration.

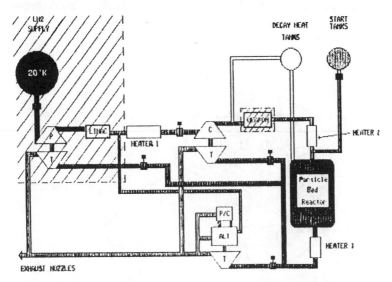

Fig. 18   Grumman open Brayton cycle system schematic.

is new and is for developing a high-power, lightweight reactor. Figure 17 shows the overall concept configuration.

Figure 18 shows the system schematic and illustrates the placement of dummy loads around the system to handle load rejection transients.

Reactor

The particle bed reactor (PBR) has a unique core in that the individual fuel assemblies are contained between a porous outer cylinder through which the coolant enters (the cold frit) and a porous inner cylinder through which the heated gas exits (the hot frit). The very small fuel particles ($500\ \mu$) allow for very rapid thermal transients and high operating temperatures. Each particle is coated with a series of materials designed to contain fission gases, resist internal pressures, and protect the particle from chemical attack or erosion by the high-temperature hydrogen. Figure 19 shows the reactor configuration, and Fig. 20 shows various aspects of the fuel and core. The thermal hydraulic behavior of this reactor requires that particular attention be paid to the pressure drops across the core and across each frit. The small fuel particle beds have a very low pressure drop, but the potential exists to have flow distribution problems, with possible local heating of the individual fuel elements, if the flow is not properly distributed and controlled along the length of the elements. Calculations by the developer show that this potential problem can be controlled by appropriate distribution of porosity along the length of the frits.

Turbine

A ten-stage turbine turning at 16,000 rpm was chosen and is shown in Fig. 21. The design techniques employed in this turbine are typical for the aerospace industry and result in a very lightweight functional turbine.

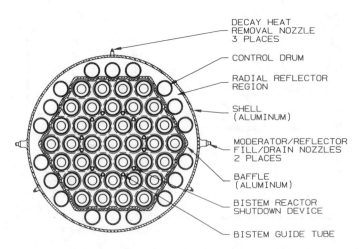

DECAY HEAT
REMOVAL NOZZLE
3 PLACES

CONTROL DRUM

RADIAL REFLECTOR
REGION

SHELL
(ALUMINUM)

MODERATOR/REFLECTOR
FILL/DRAIN NOZZLES
2 PLACES

BAFFLE
(ALUMINUM)

BISTEM REACTOR
SHUTDOWN DEVICE

BISTEM GUIDE TUBE

Fig. 19   Grumman reactor configuration.

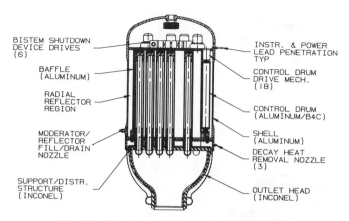

Fig. 20   Grumman reactor core configuration.

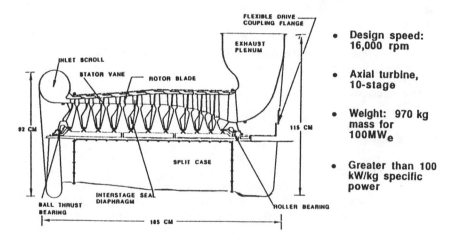

Fig. 21   Grumman turbine design.

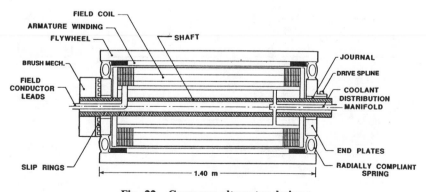

Fig. 22   Grumman alternator design.

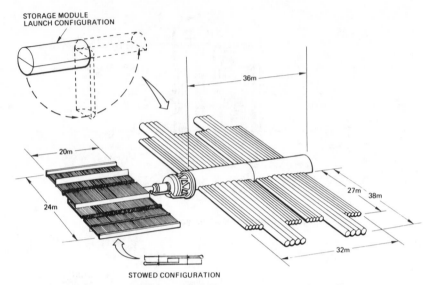

Fig. 23   General Atomic STAR-S system configuration.

Alternator
The alternator chosen has an external rotating rotor and is shown in Fig. 22.

General Atomics

System Description
General Atomics (GA) developed a closed-cycle system in the range of tens of megawatts consisting of an in-core thermionic reactor and a bank of alkaline fuel cells. The fuel cells are used to supply a burst of power, and

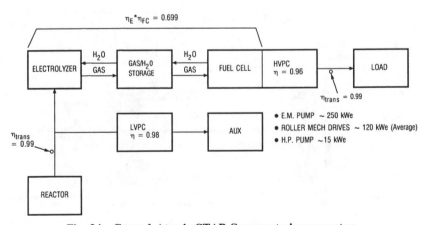

Fig. 24   General Atomic STAR-S power train parameters.

the reactor power is used to regenerate the fuel cell reactants after the power burst is completed. The water vapor effluent from the fuel cells is condensed in heat exchangers where a water coolant stream is evaporated and then collected in large, flexible radiators. The steam in the large, flexible radiators is condensed through radiation to space and then collected for electrolysis into hydrogen and oxygen for the fuel cells. As a result of closing the system, large radiators are required for operation of the reactor and the fuel cell system. Figure 23 shows the overall system configuration.

Figure 24 shows the system schematic for the reactor and fuel cell system. No power conditioning is required between the reactor system and the fuel cell system in that the reactor has a direct output of 36-V dc from the thermionic elements.

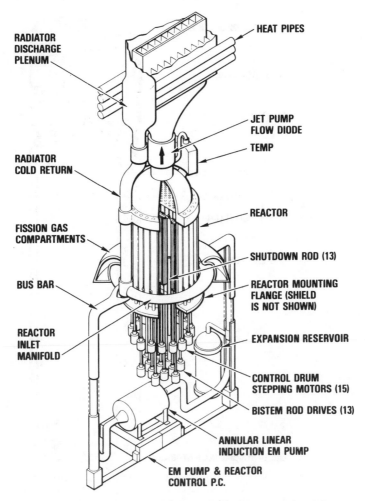

Fig. 25 General Atomic STAR-S reactor subsystem schematic.

Reactor

The General Atomics reactor is a liquid-metal-cooled, in-core thermionic reactor that generates power by virtue of the thermionic phenomenon where electrons are "boiled" off from a hot emitter and collected on a cooled collector, generating a voltage potential between the two elements. The reactor, with associated liquid metal piping, is shown in Fig. 25. Details of the thermionic core elements are illustrated in Fig. 26.

Energy Storage Device

Energy for the burst power mode is stored on the form of gaseous hydrogen and oxygen, which are processed through a number of alkaline

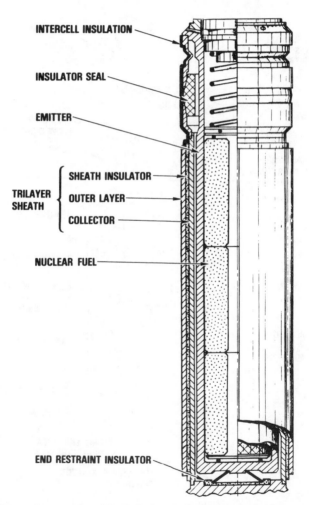

Fig. 26 General Atomic STAR-S reactor thermionic cell arrangement.

fuel cells, generating a dc current and steam. The system schematic for the fuel cells is shown in Fig. 27. The steam is collected in large, expandable bag radiators where the steam is condensed by radiation to space. The expandable radiators are rolled up during the condensation process, and the liquid water is collected in order to reinitiate the cycle. The liquid water is electrolyzed with power from the thermionic reactor and returned to the gaseous hydrogen and oxygen tanks as feed stock for the fuel cells.

Power is taken from the fuel cell banks and processed through power conditioning equipment for use on the weapon platform or other type of user.

Heat Rejection

Heat rejection from the reactor is with Nb-^{1}Zr liquid metal heat pipes operating at 1032 K. Heat rejection from the fuel cells is by cooling the water vapor effluent in large, flexible bag radiators. The bag radiators are of three different sizes and operate at 390, 403, and 453 K. Both radiator systems can be seen in an overall system layout shown in Fig. 23.

Rockwell

System Description

Rockwell developed a liquid-metal-cooled, Rankine cycle nuclear reactor system using sodium-sulfur batteries for burst power to supply power in the

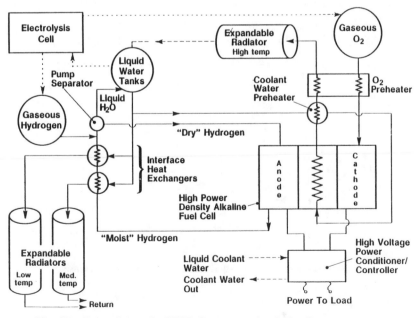

Fig. 27 General Atomic STAR-S regenerative fuel cell power system.

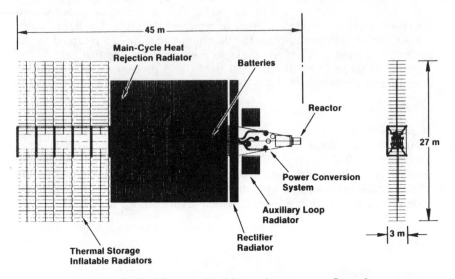

Fig. 28   Rockwell closed Rankine cycle system configuration.

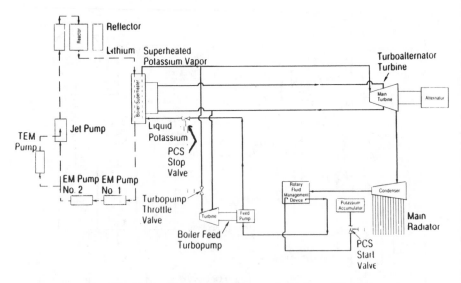

Fig. 29   Rockwell closed Rankine cycle system schematic.

tens of megawatts for short periods. The reactor system is used to recharge the battery system after the power burst is completed. The reactor system has a primary and secondary liquid metal loop, with the secondary loop being a potassium Rankine cycle driving a metal vapor turbine for power conversion. Figure 28 shows the overall system configuration.

Figure 29 shows the system schematic for the reactor and the battery system. One of four secondary loops are shown in the schematic, of which three are required to carry out the design mission.

Reactor

The reactor is a liquid-metal- (lithium) cooled fast reactor with a cermet core. Power is taken from the reactor and transferred to the secondary loops through a potassium boiler. Coolant is pumped through the system with electromagnetic (TEM) pumps. Figure 30 shows the reactor configuration and associated machinery.

Turbine

A 7-stage (4 high-pressure, 3 low-pressure) potassium vapor turbine operating at 1500 K is used in the power conversion system. The vapor is exhausted after passing the high-pressure stages, reheated, and reinjected into the low-pressure stages. A rotary fluid management device (RFMD) is

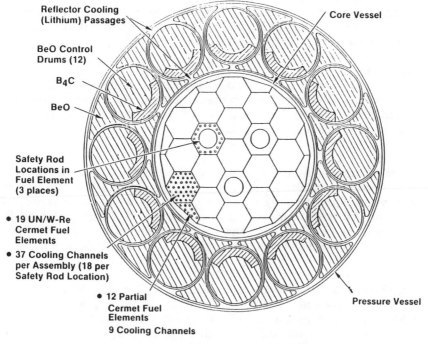

Fig. 30   Rockwell reactor configuration.

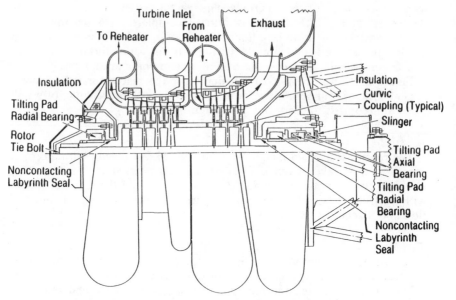

Fig. 31   **Rockwell main turboalternator turbine configuration.**

used to pump the secondary two-phase fluid in 0 *g*. The RFMD utilizes the inertia of a rotating mass of fluid to separate the two phases and pump the fluid. The turbine operates in standby at 75% of operational speed. Figure 31 shows the turbine for this application.

Generator
This design uses a wound field, round rotor, brushless-type alternator operating at 16,200 rpm. The alternator bearings are lubricated with hot potassium and lithium. The alternator is designed to operate in standby at 75% speed for the life of the system.

Heat Rejection
Heat is rejected from the secondary loop through carbon-carbon heat pipes operating at 1050 K. Auxiliary cooling loop radiators are carbon-carbon heat pipes operating at 785 K. A small amount of energy is rejected from the first-stage rectifier through a carbon-carbon heat pipe radiator using water as a working fluid. Heat rejection from the batteries is by direct radiation to space by opening panels that expose the batteries to space.

Energy Storage Device
Sodium-sulfur batteries are used to supply a short-term burst power in the tens of megawatts. The batteries are maintained at a temperature of 350°C during standby and are allowed to increase to 625°C during operation. Panels are opened during or after operation and the batteries are allowed to radiate to outer space.

Summary

The work to date on advanced space power concepts has highlighted several important areas. The first is that a tremendous amount of long-life power can be packaged in a small, relatively lightweight package, on the order of a 1000 or so kg (the reactor only). Second, power conditioning is a technology area where there is potential for great weight savings. There is a significant variance in the weights of power conditioning systems studied and proposed to date, indicating that the best solution is yet to be found. The third area highlighted is that closed-cycle systems remain heavy with respect to equivalent power open-cycle systems, primarily due to the heat-rejection systems.

The programs now under way within the Department of Energy, SDIO, and NASA are addressing these issues as they relate to specific missions, and all indications are that advanced nuclear power systems will have a very important role in future space programs.

Propulsion Systems

R. Joseph Cassady*
Rocket Research Company, Redmond, Washington

Introduction

One of the principal fluid-handling systems onboard any spacecraft is the propulsion system. Thermal-hydraulic problems are therefore of great concern to propulsion system designers. To understand the types of thermal-hydraulic problems that can occur, it is first necessary to distinguish between the various types of spacecraft propulsion systems and the environments in which these systems must operate. This chapter will give a brief introduction to each type of propulsion system and the components that comprise them. Later sections of the chapter will discuss the particular thermal-hydraulic problems each may encounter.

Types of Propulsion Systems

Spacecraft rely on propulsion systems for a variety of functions, including orbit acquisition or transfer, attitude control, and stationkeeping. Orbit acquisition is classified as primary propulsion and is usually performed by large upper stages. These upper stages perform the orbit transfer in a matter of hours, usually with engine burns of several hundred seconds duration. Attitude control and stationkeeping functions are referred to as secondary propulsion and are performed by propulsion systems integrated into the spacecraft. Because these functions require long-term storage of fluids in the microgravity environment, as well as thermal integration into the spacecraft design, these systems are more often concerned with fluid-thermal issues. The following sections describe the types of spacecraft propulsion systems currently in use or expected to be used in the near future.

*Senior Development Engineer, Electric Propulsion Technology Department.

69

Chemical Propulsion Systems

Chemical rocket propulsion systems typically combust a fuel and an oxidizer, then accelerate and discharge the heated combustion gases through a nozzle to achieve thrust. Chemical rockets include those that store the reactants in liquid form (liquid rockets) and those that store the reactants in solid form (solid rockets). From the perspective of thermal-hydraulic problems, only the liquid rocket systems are of interest. Of these, there are two categories: bipropellant rocket systems and mono-propellant rocket systems. Each of these types of liquid rockets has its own unique requirements of fluid management in space.

Bipropellant rockets store the fuel and the oxidizer separately and feed them into an injector, which mixes them in the correct proportions to achieve stable, reliable combustion. Examples of these include H_2/O_2 and MMH/N_2O_4 systems. Bipropellant rocket systems are extremely susceptible

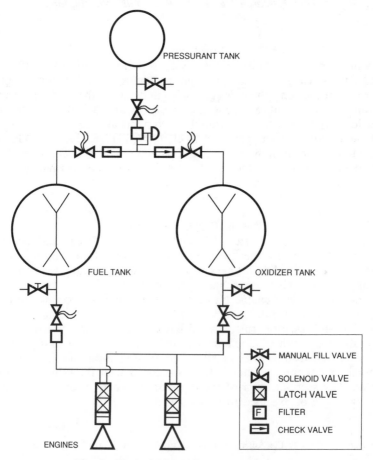

Fig. 1 Typical bipropellant propulsion system.

to thermal-hydraulic concerns since they must store and transfer two different fluids with differing physical properties. A typical propulsion system layout for a bipropellant rocket engine is shown in Fig. 1.

Monopropellant rockets utilize propellants that contain both an oxidizing agent and combustible matter in a single liquid. This liquid must be heated or catalyzed to decompose and yield hot combustion gases. A typical monopropellant is hydrazine (N_2H_4). When thermally or catalytically decomposed, hydrazine forms H_2, N_2, and NH_3 gases. A monopropellant system typically stores and feeds propellant in liquid form to a thermal reactor or catalyst bed, where the decomposition reaction occurs and the hot gases are then exhausted out via a nozzle. Figure 2 shows a diagram of a monopropellant hydrazine propulsion system.

Electric Propulsion Systems

Electric propulsion devices are rocket engines in which the energy to be converted into thrust derives principally from the coupling of electrical power into the propellant. Resistojets, arcjets, and ion thrusters are examples of electric propulsion systems. Resistojets and arcjets obtain thrust through the heating of propellant gases by electrical means and the subsequent thermal expansion of the hot gases. Both resistojets and arcjets

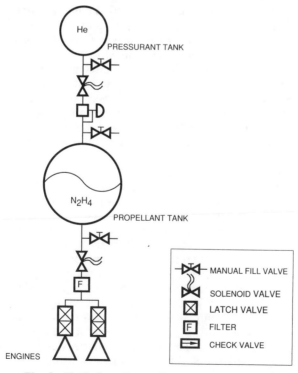

Fig. 2 Typical monopropellant propulsion system.

are capable of utilizing the type of monopropellant feed system shown previously in Fig. 2. In either of these systems, the chemical energy released during the decomposition of hydrazine increases the enthalpy of the gases, and this enthalpy is then further increased by the electrical energy. The enthalpy contribution of the chemical reaction is typically 40–50% below 1 kW and less than 15% above 1 kW. The ion engine is an electrostatic accelerator. It is an example of an engine where the only energy source is electric. Gas is introduced into a large plenum chamber, where it is ionized by electron bombardment, then accelerated out of the thruster under the action of a strong voltage gradient applied across a set of grids that form the aft wall of the chamber. The ions that are accelerated out of the chamber achieve very high exhaust velocities.

The motivation for the use of electric propulsion systems is the better utilization of the mass delivered to orbit. Use of chemical propulsion systems limits the fraction of delivered mass devoted to payload because chemical systems' specific impulse is limited to approximately 450 lbf-s/lbm by the energy available in the chemical bonds. Electric propulsion systems are not limited by this chemical bond energy and therefore achieve much higher specific impulse values. Satellite planners are now including electric thrusters in their designs for this reason. A review of the three types of electric propulsion systems now in use or under consideration for satellite applications is given in the following paragraphs.

Resistojet thrusters have been developed for many years and are now flying in a stationkeeping role on many communications satellites. A resistojet adds enthalpy to the propellant via a high-temperature coil that may be either radiatively or directly coupled to the gas flow, as illustrated in Fig. 3. Resistojets can utilize many different types of propellants, including N_2H_4, NH_3, H_2, and organic waste gases such as CO_2 and CH_4. The omnivorous nature of the resistojet, especially in regard to the latter gases, has led to its consideration as reboost propulsion for the NASA manned space station. In another similar application, the designers of the Industrial Space Facility selected resistojets using water as propellant for their reboost propulsion.

Arcjet thrusters are similar in concept to resistojets, but rather than imparting enthalpy to the gas through a resistive coil, the arcjet utilizes an electric arc discharge through the gas itself. In this way the arcjet avoids the materials limitations imposed by the coil and heat exchanger walls in the resistojet. Arcjet thrusters can operate over a wide range of power (1–100 kW) by simply scaling the physical dimensions of the device. Arcjets are therefore considered as candidates for missions ranging from stationkeeping to orbit transfer. Like the resistojet, an arcjet can operate on many different types of propellant, including N_2H_4, NH_3, and H_2. A schematic of a thermal arcjet thruster system is shown in Fig. 4.

Ion engines are in a completely separate class from chemical or electro-thermal rocket engines. In an ion engine, electrical energy is used in two ways: 1) to ionize the propellant and 2) to electrostatically accelerate the ions to very high exhaust velocities. No thermal-to-directed kinetic energy conversion takes place. Therefore, the criteria for selecting a propellant are

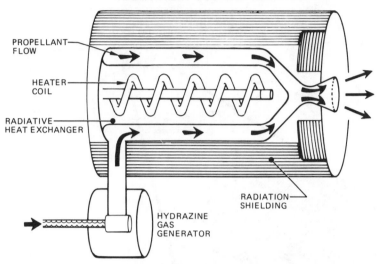

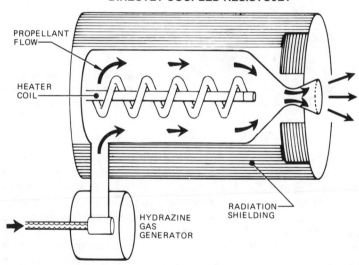

Fig. 3 Resistojet propulsion system designs. (Courtesy of Rocket Research Company.)

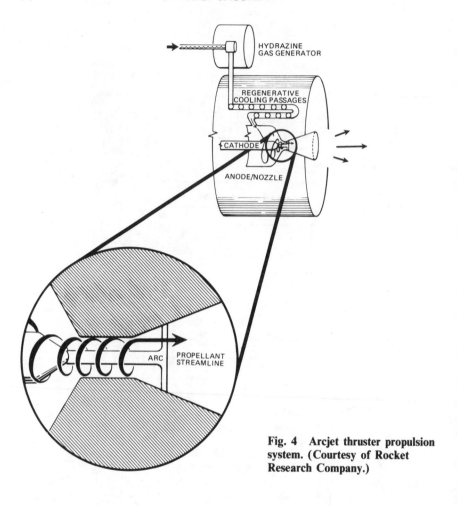

HYDRAZINE
GAS GENERATOR

REGENERATIVE
COOLING PASSAGES

CATHODE

ANODE/NOZZLE

ARC   PROPELLANT
       STREAMLINE

Fig. 4   Arcjet thruster propulsion
system. (Courtesy of Rocket
Research Company.)

completely different. An ion engine performs best with a propellant that has
a high molecular weight and a low ionization potential. Such propellants
include Hg, Xe, and Kr. Until recently, Hg was the propellant of choice.
However, environmental and spacecraft contamination concerns have virtually eliminated it as a candidate for near-Earth missions. Figure 5
illustrates the concept of the ion engine.

A unique aspect of all of these electric thrusters, compared with chemical
systems, is the requirement for the propellant to be delivered in gaseous
form. This implies some type of boiler or other means of vaporizing the
liquid propellant, or it requires gaseous storage of the propellant. For most
spacecraft applications, gaseous storage is impractical from a mass and
volume standpoint; hence, vaporizer designs must be considered. Methods
of vaporization include flash vaporization, catalytic decomposition as in a
hydrazine system, or heat addition as in a boiler.

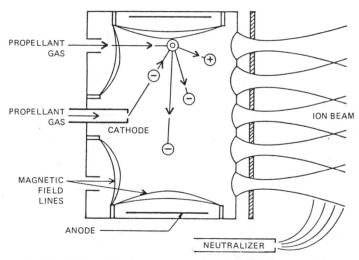

Fig. 5 Electron bombardment ion thruster propulsion system.

Propulsion System Components

Tankage and Fluid Feed

Liquid storage and transport in the microgravity environment is a difficult problem. In order to supply the propellant to the engines, some means of expulsion from the tanks is required. Large chemical upper stages may be pump fed or use pumps to initiate combustion and then utilize their own acceleration to provide positive expulsion. Spinning satellites can also achieve positive expulsion by configuring the feed system to take advantage of the centripetal acceleration about the spin axis. However, three axis-stabilized satellites must rely on some other means of removing fluid from the tanks. In the absence of gravity, the liquid does not collect or pool in one location (i.e., over the outlet) near the walls.

Various expulsion techniques have been devised, including gas pressurant, bladders, screens, and surface tension or wick devices. Figure 6 illustrates several examples of surface tension devices. Total communication trap devices utilize a liner and a liquid trap to retain liquid near the tank outlet, even under the influence of vehicle accelerations. Vanes are used in low-acceleration systems with low flow rates to induce flow of the liquid to the tank outlet. Liquid fillets form where the vanes almost touch the tank walls and where the vanes intersect. Gallery systems are applied where liquid must also be transferred to the tank outlet, but are used for higher flow rates and at higher vehicle accelerations. Galleries are rectangular cross-sectional arms with windows cut into one side. The windows resist the entry of gas until the liquid pool in the tank is depleted. A final example of a trap propellant management device (PMD) is the total control trap system. This system employs multiple parallel screens to control the

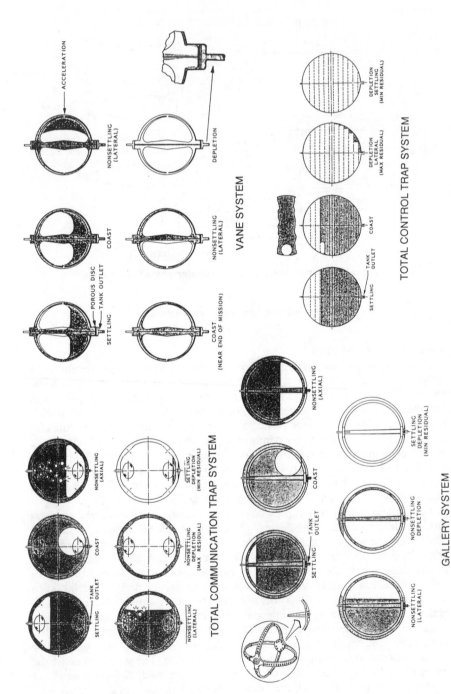

Fig. 6 Propellant management devices. (Courtesy of Lockheed Corporation.)

location of the liquid in the tank. It requires an axial acceleration to settle the propellant, but once settled maintains the liquid in contact with the walls over the tank outlet. The fundamental principle employed in all of these devices is that the inertial forces acting on the fluids in the microgravity environment can be overcome by using the viscous forces of the fluid. The advantage of these surface tension systems is their simplicity and high reliability compared to systems with active elements.

Engine Nozzles and Components

Cooling of engine components is a requirement of many spacecraft thrusters—both chemical and electric. A useful interaction between the fluid management system and the propulsion system is regenerative cooling of the rocket engine nozzles, combustion chambers, and other components. Regenerative cooling is the circulation of a propellant through passages in the rocket engine components to cool those components prior to being combusted and expelled. It is common practice on many large bipropellant rocket engines, especially those using cryogenic fuel and oxidizer combinations, such as the RL-10 engine used on the Centaur upper stage.

Electric thrusters have also been designed with regenerative cooling. In electric thrusters the regenerative cooling can be used both to cool the engine components and as a means of imparting some heat into the propellant prior to the electrical heating.

Flow Regulation

One of the most critical areas for chemical or electric thrusters is flow regulation. Chemical engines rely on feed pressure at the injector for bipropellant systems or at the catalyst bed for monopropellant thrusters to maintain stable combustion. Resistojets establish a thermal balance between the propellant and the heat exchanger that requires the flow to carry heat away from the heater components. Arcjets require stable propellant flow to maintain the arc discharge away from the walls of the thruster. Ion thruster performance is very sensitive to variations in flow rate.

Power Conditioning for Electric Propulsion Systems

With the advent of electric propulsion on spacecraft, another subsystem must be included in the thermal management of the propulsion system. This subsystem is the power conditioning or power processing electronics that accepts input power from the spacecraft power bus and outputs the correct voltage and current for the electric thruster. Power conditioning units are required for all electric propulsion devices. They vary in complexity from simple resistors in series with the thruster to complex switching power converters that step up the supply voltage and regulate the power delivered to the thruster. The design of the power conditioning can have a profound impact on the overall propulsion system waste heat budget. In fact, for high-power electric propulsion devices, such as the 30-kWe arcjet thruster, it can contribute more heat than the thruster itself. For this reason, power conditioning units (PCU's) are designed to be as efficient as practically possible. At 30 kWe input power, every percent of conversion efficiency reduces the waste heat load by 300 W.

The waste heat generated by the propulsion system must be dissipated to space via a radiator. The effectiveness with which the heat can be radiated to space depends on the radiator surface temperature raised to the fourth power. For power conditioning electronics, the maximum radiator temperature is limited by the allowable operating temperature of the electronic components within the PCU. The result is a very-low-temperature radiator (approximately 300 K maximum) and therefore a large and massive physical device.

To circumvent this problem, approaches have been proposed by some electric propulsion system designers which utilize the fluid management systems onboard spacecraft to transport and dissipate heat from the propulsion power conditioning. In these designs, the propellant is circulated through the PCU and picks up waste heat. It can then either be fed directly to the engine or pumped back into the tank. The propellant tank is then used as a radiator to dissipate the waste heat. The choice depends primarily on the cooling requirements of the PCU and the flow rate of the propellant required by the thruster.

Vaporizer

For electric propulsion systems (with the exception of those using hydrazine), some form of vaporization method must be employed. The vaporizer must produce high-quality (low liquid droplet content) vapor or gas for delivery to the thruster. System design for microgravity fluid vaporization is not well developed. Techniques that have been proposed include a cyclone boiler and flash vaporization by introducing the liquid into a plenum at very low pressure. This area in particular is difficult to accurately simulate under the influence of gravity and would benefit from experiments performed under actual microgravity conditions.

Propulsion System Thermal-Hydraulic Problems

The propulsion system designer is faced with a complex trade to determined an optimum thermal-hydraulic system design. The changing environment around the spacecraft, as well as onboard thermal events, such as thruster firings, act to continually change the thermal-hydraulic equilibrium within the system. Careful thermal modeling of the propulsion system must always be performed. Fluid resting in a line may be frozen when the exposed surface of the line views deep space instead of the sun. Improper thermal isolation can result in thermal soakback through propellant lines, which causes boiling in the lines. Either extreme can be catastrophic.

In a bipropellant system, strenuous demands are placed on the fluid management system because of the disastrous effects that an imbalance in fluid feed pressure at the injector can have on the combustion process. Effects can include excessive vibration, which can damage both the engine and other sensitive components on the spacecraft, overhot combustion, which can overheat and damage engine components, and ultimately the failure of the engine. Fluid feed pressure imbalances may be caused by improper thermal design, resulting in either boiling or freezing of the propellant in the lines prior to the injector.

Monopropellant systems, such as the hydrazine system discussed previously, are not immune from these concerns. This simple system can be adversely affected in several ways by thermal-hydraulic problems, for example, by freezing in the lines caused by a sudden reduction in pressure (i.e., the opening of a valve into a large plenum). Freezing of fluid in the lines can also result from direct exposure of the propellant line to space. Freezing can result in erratic operation or even failure of the system if the ice that forms completely blocks the feed lines. On the opposite extreme, an increase in the temperature of the feed lines can result in premature thermal decomposition of the fuel while it is still in the feed lines. The effects of thermal decomposition in the lines can range from rough combustion to damage or destruction of the system components from the migration of the flame front into the feed lines.

Electric propulsion systems are also subject to thermal-hydraulic concerns. A common characteristic of these electric propulsion systems is their requirement for substantial amounts of electrical power—from 500 W to as much as 30 kW. Because of these high power requirements, electric propulsion systems and their related power conditioning systems are especially susceptible to thermal-hydraulic problems.

If flow to a resistojet were interrupted due to freezing in lines, the thruster heat exchanger elements would overheat and fail. Safeguards, such as valve heaters and line insulation, are built into hydrazine resistojets to prevent such an occurrence.

Arcjet thruster stability is dependent on a steady supply of propellant gas. The presence of two-phase flow in the feed system of the arcjet could result in mass flow rate oscillations that would in turn cause voltage oscillations. If these oscillations grow large, excessive wear of the thruster can result, leading to reduced engine lifetime.

Ion engine flow control is critical to correct operation. Flow rate variations will result in production of doubly ionized species, which leads to higher losses. Thruster wear is also flow rate dependent. Flow oscillations will increase the rate of thruster cathode erosion.

In a fluid-cooled PCU, adverse situations that could arise include partial boiling of the fluid in the lines, leading to a loss of flow regulation and overheating of the tanks and resulting in excess pressure on the propellant or even rupture of the tanks.

Flow regulation can be upset by the formation of vapor in fluid lines. If an active feedback system is employed, vapor in the lines may result in the sensing element demanding an increase in fluid supply when a bubble passes through. Such conditions can result in bubble collapse and sudden "water hammer" pressure impulses within the propellant feed lines.

In general, the treatment of thermal-hydraulic problems in propulsion system design involves careful thermal analysis of the expected propulsion system design and its layout on the spacecraft. Based on the results of this analysis and thermal vacuum testing of high-fidelity mock-ups of the systems, conservative steps are taken to ensure that the propellant will not approach either extreme in temperature. These include insulation, thermal louvers, and heaters. These design solutions often add mass and complexity to the propulsion system, but they are necessary to ensure reliability of operation.

From the discussion presented in this section, it is apparent that the design of spacecraft propulsion systems can benefit from research into thermal-hydraulics in the space environment. Such research can impact many areas, including expulsion system design, vaporizer design, thruster regenerative cooling design, and system layout and thermal design in general. Present and future propulsion systems will benefit from this research through reductions in mass and system complexity. The problems of spacecraft propulsion system thermal-hydraulic design, now solved through conservative design, may be avoided with a better understanding of the behavior of liquids and two-phase flows in the microgravity environment.

Chapter 2. Space Station Two-Phase Thermal Management

Space Station Mechanical Pumped Loop

Richard Brown* and Joseph Alario†
Grumman Space Systems, Bethpage, New York

Introduction

For large systems such as the central thermal control system for the Space Station, capillary pumped loops are not practical because of high heat loads and long line lengths, and mechanically pumped loops are the preferred alternative. In these systems, fluid is circulated via a mechanical pump rather than by capillary forces. A mechanically pumped system contains a heat-acquisition section, consisting of a number of evaporators; a heat-transport section consisting of a pumping device, an accumulator, a means of set-point control, and perhaps a noncondensable gas trap; and a heat-rejection section made up of multiple condensers, each of which rejects heat to heat pipe radiators. There are two generic types of mechanically pumped systems currently under development for potential use on the Space Station: separated flow and mixed two-phase flow, as discussed later.

In the mixed two-phase flow systems, liquid is pumped into the evaporators at a constant rate, regardless of heat load. Enough of the liquid will evaporate to match the heat load, and a mixture of liquid and vapor will exit from each evaporator and flow to the condensers in a two-phase return line. At high heat loads, this mixture will be of high quality, whereas at low heat loads, the quality will be low and the two-phase return line will contain mostly liquid. The drawbacks to mixed systems are the high liquid inventory that must be carried to fill up the two-phase return line under conditions of low heat load, as well as the higher pressure drops associated with two-phase flow. Although the knowledge base of two-phase flow in 0 *g*

*Project Engineer, Thermal Systems.
†Manager, Thermal Systems.

is limited, some initial experiments[1,2] have indicated that two-phase flow pressure drops in 0 g can be three times higher than those in 1 g.

The separated concepts take their name from the fact that the long transport lines contain strictly liquid or vapor, with no mixing of the two phases. This approach minimizes overall system pressure drop and fluid inventory by avoiding two-phase flow effects. In order to accomplish this, however, some means of either controlling the flow rate of liquid into the evaporators to match the heat load or of separating the excess liquid flow from evaporator feed and returning it to the pump is required. Thus, the separated systems trade off some additional complexity in order to avoid the higher pressure drops and uncertainties of two-phase flow in 0 g.

Space Station Thermal Bus Design Requirements

Although not all of the requirements have been definitized for the Space Station thermal control system (TCS), the following list describes some of the more basic requirements and design goals for the thermal bus:

1) Heat loads: The initial station will require that the TCS be able to dissipate about 75 kW of power.

2) Growth: The TCS should have the ability to grow to accommodate anticipated future heat loads of up to 300 kW.

3) Set point: Two loops are envisioned, one operating at a set point of 70°F for equipment cooling and one at 35°F for cabin humidity and metabolic load control. Each loop should be able to be reset to the other temperature in case of failure.

4) Isothermality: The specified isothermal band in which the system should operate at a given set point is 9°F; however, smaller isothermal bands are highly desirable. The isothermality must be maintained under all conditions of load and orbital environment.

5) Reliability: With a required 30-yr life, the TCS must be very reliable. Components should be designed so that they can be maintained on orbit.

6) Power consumption: Power consumption must be minimized.

7) Heat acquisition: The TCS must accommodate a variety of heat source interfaces, including cold plates for equipment mounting and single- and two-phase fluid loops. The temperature difference (ΔT) between the heat source and the set-point temperature should be minimized. Anticipated heat fluxes are on the order of 1 W/cm².

8) Heat rejection: Efficient heat transfer between condensers and radiators is essential because any extra temperature drops will increase the required number of radiators.

9) Predictability: The system should be amenable to analytical modeling.

Separated-Flow Systems

Two separated-flow systems have been proposed for the Space Station Central TCS. These systems, under development by Grumman Space Systems Division and Lockheed Missiles and Space Company are described in the following subsections.

Grumman Thermal Bus Concept

The Grumman separated thermal bus concept is a separated two-phase system using ammonia as the working fluid. Figure 1 is a schematic of the basic concept. Liquid is pumped via a constant-speed, positive displacement pump to each of a number of evaporators that are plumbed in parallel. Upstream of each evaporator is a reservoir and an on/off liquid supply valve. Ultrasonic sensors mounted on the reservoir sense the fluid inventory in the reservoir. When the inventory is low, the sensor opens the valve for a predetermined period of time to allow pressurized liquid from the pump to fill the reservoir, at which point the valve closes. Capillary forces generated in the evaporator cause liquid to flow into the evaporator as needed to match the heat load. This liquid evaporates, and only vapor leaves the evaporator. In this way, phase separation is maintained in the transport lines. Depending on the condition (open/close) of the liquid supply valves, excess flow circulates through a pump bypass line containing a passive relief valve that serves to keep the pump outlet pressure relatively constant.

The vapor exiting the evaporators flows into a main vapor line and continues on to the condensers, where it is condensed as heat is transferred to mechanically attached heat pipe radiators whose function is to reject the waste heat to space.

Temperature control of the bus is maintained by liquid blockage of the condensers. Varying the size of the flooded region in the condenser effectively changes the thermal conductance between the bus and the heat sink (radiators). This allows the bus to operate over a narrow temperature range despite changing heat loads and environmental sink temperatures. As either the heat load or the sink temperature is decreased, excess liquid flows into the condensers from the accumulator, reducing the thermal coupling between the condensers and the radiators as the heat-transfer area is blocked, thereby maintaining a constant bus temperature. If the heat load or the environment temperature increases, liquid flows back into the accumulator from the condensers, opening up the heat-transfer area and allowing the increased load to be rejected at the same bus temperature.

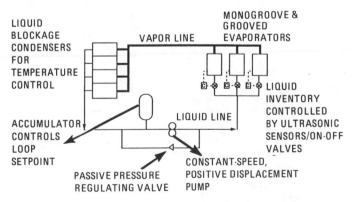

Fig. 1   Grumman separated-flow thermal bus concept.

The function of the accumulator, then, is to expel liquid from the accumulator into the condensers when a greater degree of flooding is required and to draw liquid into the accumulator when less flooding is required. Several different types of accumulators can be used to accomplish condenser flooding, including a metal bellows with nitrogen pressurant on one side and ammonia on the other, an electromechanical piston accumulator, and a two-phase accumulator that contains no moving parts. Although all three have been successfully tested, the two-phase accumulator is the preferred approach and will be discussed in more detail later.

Grumman Prototype Thermal Bus System

Grumman, under contract to NASA-JSC has built, delivered, and tested two thermal bus systems. The first, the thermal bus technology demonstrator (TBTD),[3] verified the concept during ambient and thermal/vacuum testing in 1985 at NASA-JSC. Based on the success of the TBTD, a second higher fidelity system was constructed (also under contract to NASA-JSC). This system, the Grumman prototype thermal bus system (GPTBS)[4,5] underwent ambient testing at NASA-JSC in January 1989. A description of the GPTBS and pertinent test results follows.

The GPTBS is a nominal 25-kW system using ammonia as the working fluid. The heat-acquisition section consists of five different evaporators, each designed to simulate a different type of heat-acquisition interface. Four of the evaporators utilize a number of legs of monogroove heat pipe extrusion welded together at the flanges and connected by liquid and vapor headers. A schematic of a typical monogroove evaporator is shown in Fig. 2. Liquid is drawn into the evaporator from the reservoir by capillary forces generated by the monogroove slot and fine circumferential grooves

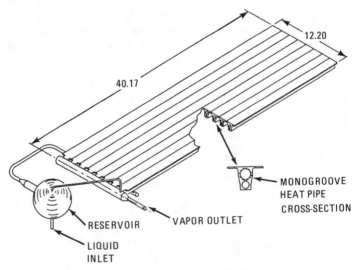

Fig. 2   Monogroove cold plate.

in the vapor channel. Evaporation occurs from the fine grooves and single-phase vapor exits the evaporator.

The advantages of the monogroove evaporator is that it is relatively easy to manufacture since the monogroove legs are an extrusion, it is a good pressure vessel, and it is based on existing technology that was developed for the space constructible radiator[6] program and currently is utilized on the space erectable radiator program,[7] discussed in the section on Central Heat Pipe Radiator System. Two of the four monogroove evaporators contain redundant fluid passages. Both of these and one other use electrical heaters as simulated heat sources, whereas the fourth is heated via an integral single-phase water/glycol heat exchanger.

A second type of evaporator is a grooved flat-plate evaporator, shown in Fig. 3. This is essentially a flat version of the monogroove design in which fine grooves are machined on the underside of the planar heat-transfer surface. As a result of the more efficient arrangement between the grooves and the heat source, this design has a somewhat higher heat flux capability than the monogroove design as well as an increased conductance. The drawback of the grooved evaporator is that it is more complex to manufacture than the monogroove evaporator. Electrical heaters simulate the heat source for the grooved evaporator.

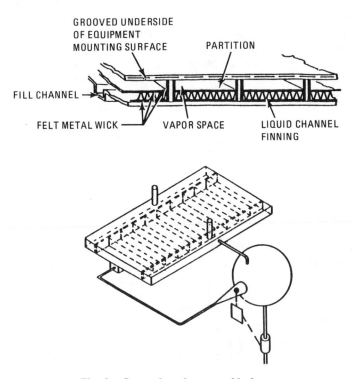

Fig. 3   Grooved equipment cold plate.

Each evaporator has a reservoir and a liquid supply valve located upstream. The spherical reservoirs are sized to provide an approximately 5-min supply of liquid in $0\,g$ at full evaporator heat load. The liquid supply valve, therefore, would only have to cycle once every 5 min at the maximum heat load. With a projected life of 10 million cycles (1 million cycles have been demonstrated), the life expectancy would be close to 100 yr.

Mounted on the reservoir liquid exit is an ultrasonic liquid presence sensor, consisting of a transmitter crystal and a receiver crystal. When the reservoir has been depleted of liquid so that vapor exists between the sensor crystals, the sensors sense a "dry" condition and command the liquid supply valve to open for a preset amount of time, enough to fill the reservoir without flooding the evaporator.

To ensure that fluid flow into the evaporators is by capillary rather than gravity forces, the ultrasonic sensors are positioned so the liquid level in the reservoir cycles between the monogroove slot level and 0.25 in. below that level. This is similar to testing a heat pipe at adverse tilt, and although it confirms capillary operation of the evaporator, it limits the $1\text{-}g$ cycle time to 1 min or less because of the limited fluid displacement.

The transport section consists of liquid and vapor lines approximately 200 ft long, redundant pumps, redundant filters, accumulator, and liquid and vapor noncondensable gas traps. The pump is a positive displacement, magnetically coupled gear pump that runs at a constant speed of about 2700 rpm, putting out about 1.2 g/m of ammonia flow. This type of pump requires virtually no subcooling and can easily pass a mixture of liquid and vapor through it. A simple, spring-loaded relief valve in the pump bypass line maintains the pump outlet pressure at about 15 lbs per sq. in. difference (psid) higher than the system saturation pressure.

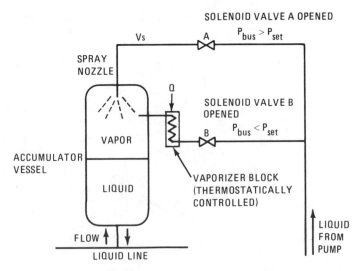

Fig. 4 Grumman prototype TBS accumulator.

As mentioned earlier, set-point control of the thermal bus is maintained by liquid blockage of the condensers. A two-phase accumulator is used to accomplish this. The accumulator operates on the principle of direct energy transfer into and out of the vapor space of the accumulator via subcooled liquid spray (direct contact condensation) and warm vapor addition from an external vaporizer block (stored thermal energy). Illustrated in Fig. 4, the accumulator system consists of a tank, a vaporizer block, and one three-way or two two-way solenoid valves. Note that all active components are external to the accumulator tank for easy access if maintenance is required. If the bus set-point pressure rises above its prescribed value,

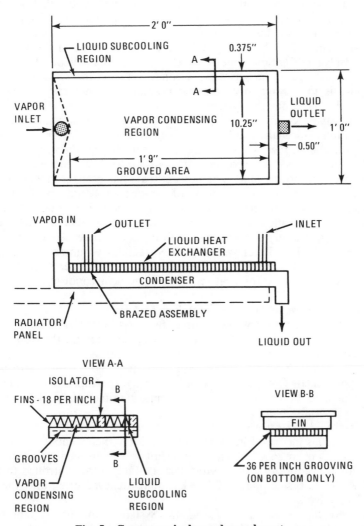

Fig. 5   Grumman single condenser layout.

solenoid valve "A" is opened to allow subcooled liquid to flow through a spray nozzle and into the accumulator. Vapor will condense on the liquid droplets and cause a partial collapse of the vapor space in the accumulator, thereby reducing the pressure in the accumulator and causing liquid to flow from the blocked zone of the condensers into the accumulator. This effectively reduces the system pressure. On the other hand, if system pressure falls below its prescribed value, solenoid valve "B" is opened, permitting a small amount of liquid to flow into the vaporizer block, where it is vaporized in coiled tubing within the block. The vapor then enters the accumulator, increasing the pressure and causing liquid to flow into the condensers, blocking area and increasing system pressure back to its nominal level. Heaters imbedded in the vaporizer block are thermostatically controlled to keep the block at a temperature high enough to permit vaporization of the liquid stream.

Two noncondensable gas (NCG) traps are included in the GPTBS for purposes of evaluation. One is located on the main vapor line, and the other on the liquid line between the condensers and the accumulator. Their purpose is to remove any noncondensable gases that may be generated within the bus from the system.

The heat-rejection section is made up of two banks of condensers, single and double. Internally, both types of condensers are similar; each consists of a condensing region surrounded by a subcooling region (Fig. 5), with each region finned to provide for high thermal conductance from the condenser wall to the fluid. The bottom plate has grooves machined into it to provide a path for the flow of condensed liquid from the vapor condensing region to the liquid subcooling region. The double condensers have only one vapor inlet and liquid outlet port, thereby reducing the number of plumbing connections. Both types of condensers can mate with a space erectable radiator system (SERS) radiator on its bottom surface. A finned fluid heat exchanger has been bonded to the upper surface of every condenser to provide for heat removal in ambient conditions.

Grumman Separated Flow Thermal Bus Test Data

The Grumman prototype thermal bus was tested in ambient during January 1989 at NASA-JSC.[8] The following summarizes the results of the ambient test.

Thermal Bus Startup Sequence. Both hot and cold startups were successfully performed. Cold startup sequence was very simple, with a heat load being able to be applied to the bus almost immediately. Set point was achieved within 25 min.

Transient Load Effects on System Performance. At a 70°F set point, the heat load was increased in approximately 5-kW increments from an initial value of 6 kW to 26 kW. System set point was controlled to within 1.5°F. Individual evaporator vapor outlet temperatures were controlled to within 2°F. At 35°F, system loads of 12.8 kW were achieved using only the single condenser bank.

System 10:1 Heat Load Turndown. System turndown capability was successfully demonstrated at 70°F by reducing the system heat load from

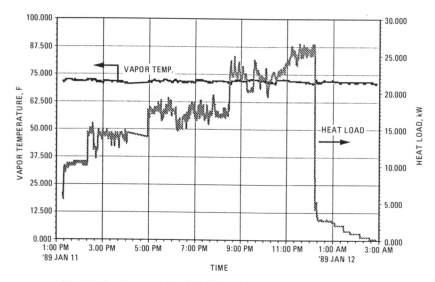

Fig. 6   System transient loads/turndown effects/quiescent load.

26 kW to 2.5 kW in a single step. System set point was tightly maintained, and there were no noticeable effects on system behavior.

Quiescent Load Operation. Stable system operation at a total heat load of 250 W (100:1 turndown) was demonstrated, with the bus set point remaining constant within 1°F. Lower loads could not be attempted due to facility limitations. As a result of ambient heat leaks at the 35°F set point, the minimum heat load that was achieved was 2.5 kW.

Figure 6 shows system vapor temperature response to changes in heat load during the previously mentioned three test series.

System Set-Point Temperature Control. The two-phase accumulator in conjunction with the liquid-blocked condensers worked extremely well in controlling system set point throughout the test. Both the vapor temperature and the individual evaporator temperatures were maintained within 2°F. Vapor line pressure was held to within about 5 psia.

System Subcooling Requirements. The system subcooling requirements are defined as the minimum difference between the set-point temperature and the pump inlet temperature that permits stable operation. For the 70°F set point, 2.5°F subcooling was required for a load of 5 kW and 7.2°F for a load of 22 kW. At a 35°F set point, about 3–5°F was required at the maximum tested load of 12.8 kW. These system limits represent the minimum amount of subcooling required by the evaporators.

System Response to Variable Heat Sink Environments. Five simulated orbits were run to determine system response to changing environmental sink conditions. During the first orbit the heat load was 5 kW, during the second and third it was 25 kW, and during the fourth and fifth orbits the load was varied between 5 and 24 kW. Throughout this test, in which the sink temperature ranged from 15 to 45°F, the set point was controlled to

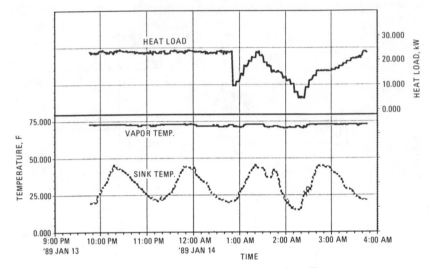

Fig. 7 **Variable environment: orbital profiles.**

within 1.5°F. Figure 7 illustrates set-point control for the latter four simulated orbits.

Heat Load Sharing. The ability of an evaporator to load share, i.e., transfer heat to a payload or fluid heat exchanger in order to keep it warm, was demonstrated using the single-phase water heat exchanger. Heat load sharing of 3.6 kW, or 40% of the system heat load, was achieved.

System Pressure Drop. At the maximum heat load of 26 kW, the pressure drop across the entire vapor line was 0.1 psid.

Pump Electrical Power Usage. During the course of the test, the positive displacement pump consumed approximately 37 W of power.

System Transient Response During Set-Point Changes. The Grumman bus exhibited the ability to execute a set-point change from 70 to 35°F and also from 35 to 70°F. The time required to reach the new set point is dependent on the ambient environment, system heat load, and condenser sink temperature.

System Control Complexity. Automatic control of the Grumman bus, including evaporator flow control and system set-point control, were demonstrated. To change set point, the new pressure is merely input to the system controller through a keypad. The ability of an evaporator to recover from a dryout without removing the heat load was demonstrated manually, since the controller was not programmed to have this capability. Also, a rise in the pressure drop across the system filter from 3 psid to over 10 psid was passively accommodated by the system.

System NCG Sensitivity, Collection, and Venting Capability. At the 35°F set point approximately 10 in.[3] of NCG was injected into the bus upstream of the grooved cold-plate reservoir, and at the 70°F set point approximately 34 in.[3] of NCG was injected. No effects on system operation were noticed.

In addition, no problems occurred during the set-point reduction. During this test, neither gas trap indicated the presence of NCG in the trap.

Evaporator Load Biasing. All four of the electrically heated evaporators demonstrated the ability to operate with nonuniform heat loads. No noticeable effects were observed on system performance.

Condenser Design Capacity. The heat-rejection capacities of both the single and double condensers were determined at the 70°F set point. At a sink temperature of 35°F, the single condensers were able to reject 19 kW, and the double condensers rejected 13 kW. The diminished capacity of the double condensers was a result of the internal routing of the liquid outlet line in the double condensers, which caused reheating of the subcooled liquid.

Evaporator and Heat-Exchanger Performance. The following are the maximum heat loads and fluxes that were obtained on the various evaporators:

$$6.3 \text{ kW} \ (2.26 \text{ W/cm}^2)$$

$$9.3 \text{ kW} \ (0.75 \text{ W/cm}^2)$$

$$9.8 \text{ kW} \ (1.27 \text{ W/cm}^2)$$

$$2.4 \text{ kW} \ (1.78 \text{ W/cm}^2)$$

$$3.5 \text{ kW} \ (0.84 \text{ W/cm}^2)$$

Lockheed Thermal Bus Concept

A second type of separated-flow thermal bus concept, illustrated in Fig. 8, has been proposed by Lockheed Missiles and Space Company (LMSC). In contrast to the Grumman concept, phase separation in the evaporators is controlled by a complex wick rather than by active flow control. Since the flow to each evaporator is constant, an extra recirculation line is required to transport the excess liquid from the evaporators back to the pump inlet. The system operates as follows.

Liquid ammonia is supplied to each of a number of parallel heat-acquisition devices (HAD's) via a centrifugal pump. An orifice at the entrance to each HAD serves to keep the flow rate relatively constant. Each HAD contains a wicking system (described in more detail later) that is designed to transport to the evaporation sites by capillary forces only enough liquid necessary to match the heat load. At the maximum heat load, approximately 50% of the supply liquid will evaporate with the remaining drawn into a return tube by the pump. The liquid in the return line from each HAD passes through a heat exchanger in which it is cooled by the supply liquid in order to ensure that the return liquid entering the pump is adequately subcooled. From there, the return liquid flows through a recirculation liquid pressure regulating valve (LPRV), which prevents two-phase flow in the return line by maintaining upstream pressure and back to the pump inlet. At low heat loads, a solenoid valve is opened to

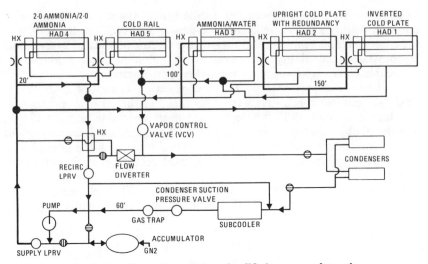

Fig. 8   **Lockheed thermal bus simplified system schematic.**

divert some of the return liquid through a preset metering valve into the vapor line and to the condensers in order to gain sufficient subcooling. The setting of the metering valve is different for different set points.

Vapor generated in the HAD's is transported in a main vapor line to the condensers, first passing through a motor-driven vapor pressure regulating valve. Two types of set-point control schemes are possible. In the first, liquid blockage of the condensers is used to control the amount of surface area where condensation takes place. At high heat loads and/or warm sink conditions, the blocked zone will be small, whereas at low loads and/or cold sink conditions, it will be large. Liquid blockage does not require the use of the vapor pressure regulating valve, and the recirculation liquid pressure regulating valve can be replaced by a fixed orifice. The second control scheme involves the use of nitrogen gas to block a portion of the condenser area. Here, both of the aforementioned control valves are needed, as well as a means of controlling the amount of gas that enters the condenser.

After leaving the condenser, saturated liquid ammonia enters the sub-cooler, which condenses any remaining ammonia vapor and provides the required subcooled liquid ammonia to the pump and HAD's. The liquid passes through a motor-driven liquid pressure regulating valve, which maintains upstream pressure to prevent vapor ingestion in the liquid return tube. This valve can be replaced by a fixed orifice if gas-blocked condensers are utilized. Next in line is a noncondensable gas trap, whose function it is to remove any noncondensable gas that may be in the liquid stream. Upon exiting the gas trap, the liquid flows through a filter, past a bellows-type accumulator and into the pump inlet. The accumulator is used to establish system operating pressure, dampen pressure transients in the system, and accommodate liquid inventory changes. The backside of the accumulator contains nitrogen, whose pressure is controlled to the desired system operating pressure. Fluid leaving the pump travels through an 0.2-μ filter and a motor-driven supply liquid pressure regulating valve. This valve maintains downstream pressure to the HAD's.

LMSC Thermal Bus Ground Test Unit

In support of space station development, Lockheed has built a thermal bus ground test unit (GTU) and provided it to NASA-JSC for evaluation in the thermal test bed.[9] The heat-acquisition section consists of five HAD's plumbed in parallel: two are cold plates, one is a cold rail, one is a single-phase water/two-phase ammonia heat exchanger, and the fifth is a two-phase ammonia/two-phase ammonia heat exchanger. All HAD's are of the same basic design consisting of multiple evaporator tubes in parallel. Figure 9 shows the evaporator tube/wick assembly, and Fig. 10 is a schematic of a cold plate. A saddle and wave spring keep the porous tube/wick assembly in contact with the circumferentially threaded tubes. The porous tube/wick assembly is made up of supply and return liquid channels that are formed from 1-μ porous nickel press fit into a 220-μ porous nickel U cup. Liquid is fed from the supply channel, through first the 1-μ wick and then the 220-μ wick and into the grooves by capillary forces. Evaporation occurs from the grooves. Excess liquid not needed for evaporation (50% at maximum heat load) is drawn into the return line by the pump.

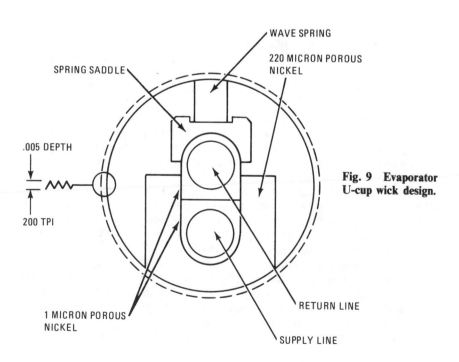

Fig. 9 Evaporator U-cup wick design.

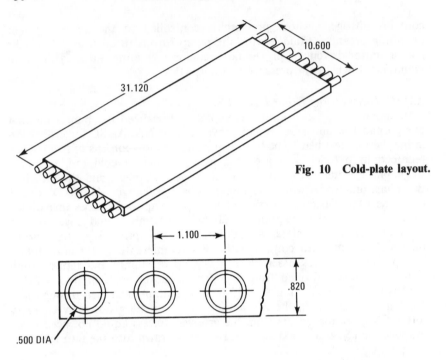

Fig. 10   Cold-plate layout.

The accumulator is a metal bellows tank with ammonia liquid on one side and nitrogen gas on the other side. In order to prevent damage to the bellows, the pressure differential cannot exceed 25 psid in either direction. This requires that relief valves be provided on both sides of the accumulator. Nitrogen pressure is controlled by the system controller to maintain the loop set point as well as to prevent overpressurization of the bellows.

The liquid-blocked condenser consists of many small cross-sectioned channels that are distributed across the width and length of the condenser. A vapor manifold provides uniform vapor flow to the channels at the inlet end, and a liquid manifold provides for the collection of condensate at the outlet end. The condenser has been designed to interface with four radiator panels in series. The amount of liquid in the condenser is controlled with the accumulator and the liquid pressure regulating valve located between the accumulator and the condensers. Downstream of the condenser is the subcooler. It is of the same design as the condenser and is used to provide for HAD and pump subcooling requirements.

A gas-blocked condenser, able to accommodate four radiator panels in series, has also been provided for performance comparison with the liquid-blocked condensers. It consists of 20–0.75-in.-diam parallel tubes spaced 1 in. apart, each made up of a single 1-μ porous tube in a 100-threads/in. threaded tube with a broached groove under the porous tube (see Fig. 11). A saddle and wave spring keep the porous tube in contact with the

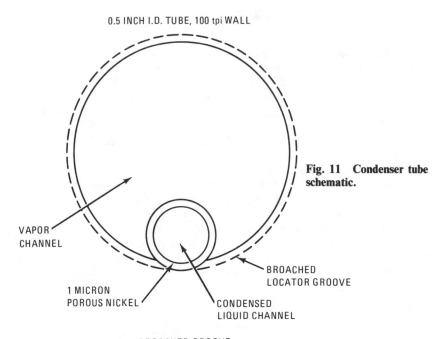

0.5 INCH I.D. TUBE, 100 tpi WALL

Fig. 11   Condenser tube schematic.

VAPOR CHANNEL

BROACHED LOCATOR GROOVE

1 MICRON POROUS NICKEL

CONDENSED LIQUID CHANNEL

BROACHED GROOVE

threaded tube. Vapor enters the condenser tubes, condenses in the threaded grooves, and flows by capillary forces into the broached groove. From the broached groove, liquid is drawn into the return tube by the pump.

LMSC Thermal Bus Test Data

As of this writing, the LMSC GTU has been delivered to NASA-JSC and is being prepared for ambient testing in March 1989. An earlier unit called the thermal bus engineering development unit (EDU) was tested at NASA-JSC in May–June 1987.[10] Although the test did verify individual HAD and condenser operation, there were a number of deficiencies cited by NASA:

1) inability to control thermal bus operation;

2) probable need for subcooling evaporator liquid return, possibly requiring an extra line across the radiator gimbal as well as extra radiator panel(s);

3) high sensitivity of bus to noncondensable gas;

4) sensitivity of bus to contamination due to small passages in control valves and 1-μ sintered wicks in evaporators and condensers; and

5) limited ability to vent noncondensable gases.

These deficiencies have been addressed in the design of the GTU.

Mixed-Flow Systems

Boeing/Sundstrand Thermal Bus Concept

The Boeing/Sundstrand thermal bus concept is a mixed two-phase flow system. Shown schematically in Fig. 12, it consists of a number of forced-flow evaporators plumbed in parallel, each with a cavitating venturi at its inlet; shear-flow or Gregorig-grooved condensers for heat rejection; a rotating fluid management device (RFMD) to provide the pumping force via pitot pumps, to separate liquid and vapor using centrifugal forces generated by rotation, and to act as a regenerator; a bellows accumulator; and a backpressure regulating valve (BPRV).

System operation is as follows. The RFMD, which is essentially a rotating drum, pumps liquid ammonia via a pitot pump through a cavitating venturi to each evaporator (see Fig. 13 for a description of RFMD operation). The cavitating venturi serves to provide a constant flow rate to each evaporator regardless of downstream pressure, providing the upstream liquid temperature remains relatively constant. The cavitating venturi's throat area is precisely sized so that liquid is accelerated through the

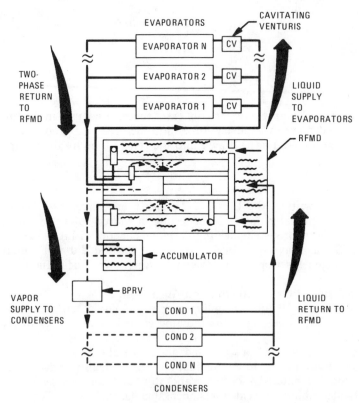

Fig. 12   Boeing/Sundstrand thermal bus simplified schematic.

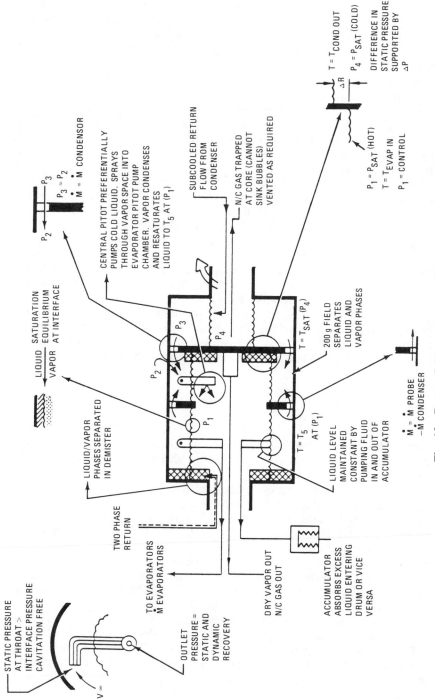

Fig. 13 Rotary fluid management device.

venturi until its vapor pressure is achieved, at which point cavitation occurs. The vapor is then recondensed in the venturi diffuser with low pressure loss. Flow rate is proportional to upstream pressure and local vapor pressure at the inlet. The latter implies that the flow rate will fluctuate if the inlet temperature varies. Since the RFMD acts as a regenerator and reheats subcooled liquid to near the saturation temperature, these fluctuations should be small.

Heat applied to the evaporators vaporizes a fraction of the fluid. The resulting two-phase mixture exiting the evaporator returns to the main chamber of the RFMD where centrifugal forces separate it into liquid and vapor. Some of the vapor passes out of the RFMD and through the BPRV to the condensers where it is condensed. Condensed liquid returns to the cold end of the RFMD, which is separated from the main chamber by a thermal barrier. Liquid level differences between the cold and warm chambers causes cold liquid to flow around the thermal barrier through several flow passages and mix with the main chamber fluid. This relatively cold liquid is pumped by the main chamber pitot pump (one of three pitot pumps in the RFMD) through atomizers into the vapor space, where it encounters the incoming evaporator two-phase flow. The atomized liquid becomes saturated by condensing some of the vapor, which then rejoins the annular rotating liquid.

Temperature control is accomplished by control of the RFMD core pressure. The BPRV regulates this pressure by balancing the upstream pressure with an internal servo-pressure. A metal bellows accumulator is provided to accommodate liquid inventory changes between the condenser and evaporator loops.

Boeing Prototype Thermal Bus System

Boeing, under contract to NASA-JSC, built and delivered a 25-kW thermal bus loop.[11] The Boeing prototype thermal bus system (BPTBS) underwent both ambient (May 1988) and thermal/vacuum testing (July 1988) at NASA-JSC.[12,13]

The heat-acquisition section consists of five types of evaporators, each representing a different thermal interface on the Space Station. These include a single-phase water heat exchanger, a two-phase water heat exchanger, a two-phase ammonia heat exchanger, a cold plate, and a cold rail. Upstream of each evaporator is a cavitating venturi designed to provide a constant liquid flow rate to the evaporator.

Both the single-phase water and two-phase ammonia heat exchangers are of a similar tube in tube design, with Fig. 14 illustrating the single-phase water heat exchanger. The source fluid flows in a finned annulus concurrently with ammonia in the grooved core. A number of parallel tubes are used, connected by inlet liquid and outlet two-phase headers.

The cold plate, shown in Fig. 15, is made up of dual coils brazed to a common mounting plate. Bolt holes are provided in the mounting plate for attachment of payloads, for test purposes, electrical heaters were used to supply heat.

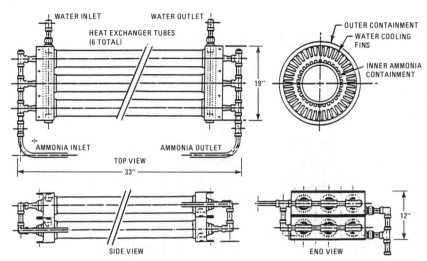

Fig. 14 The 5-kW single-phase water interface heat-exchanger design.

The transport section contains the RFMD, BPRV, filters, transport lines, and dual accumulators. The RFMD is constructed of stainless steel, aluminum, and fluoroplastic. It has a stationary hermetic housing with fixed pitot pump probes and a rotating inner drum. The BPRV, shown in Fig. 16, is constructed of stainless steel. Each of the accumulators is made up of a tank with a welded metal bellows inside that serves to separate liquid and vapor and is wrapped by thermal conditioning coils. System liquid inventory changes in the condenser and evaporator loops are accommodated by the accumulators. If more liquid is leaving the RFMD than is returning to it, the level in the RFMD will move radially outward and the bellows spring force will cause the bellows to stroke, forcing liquid out of the accumulators and into the RFMD.

Two different condenser concepts were used to demonstrate heat rejection on the BPTBS. The twin condensers, shown in Fig. 17, are designed to mate with two radiator panels. They consist of a number of parallel sculptured, extruded surfaces called Gregorig grooves that serve as condensing passages. The Gregorig groove profile is intended to maintain a thin liquid film on the condensing surfaces. Flow is directed axially along each radiator. The shear-flow condenser mates with six radiator panels. Here, the vapor flow is directed across the six radiators, with the vapor entering at the wide end of a number of parallel, tapered grooves in which a constant vapor velocity is maintained during the condensation process. With this configuration, under less than full heat load conditions, the radiators at the exit of the condenser will reject less load than radiators near the condenser entrance.

Boeing/Sundstrand Thermal Bus Test Data
The Boeing/Sundstrand prototype thermal bus was tested in ambient and thermal/vacuum conditions during the spring and summer of 1988 at

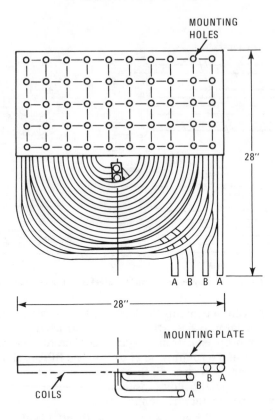

- DUAL COIL SWIRL FLOW EVAPORATOR
- HEAT LOAD SUPPLIED BY SURFACE HEATERS
- MAXIMUM EXIT QUALITY = 80%
- PRESSURE DROP < 1 PSI
- COILS BRAZED TO MOUNTING PLATE
- MATERIAL – ALUMINUM 6061-T6

Fig. 15   The 5-kW cold-plate design (swirl flow evaporator).

NASA-JSC. The following summarizes the results of the ambient test, taken from Refs. 12 and 13.

Thermal Bus Startup Sequence. The startup sequence was successfully demonstrated with both hot and cold system initial conditions. During the cold system startup, the system stabilized at the desired set point approximately 15 min after initiation of the startup procedure. Initial RFMD power consumption was 1000 W due to flooding-induced drag. Once the system stabilized, RFMD power consumption leveled off at about 520 W.

Transient Load Effects on System Performance. At a 70°F set point, as the evaporator heat loads were stepped up from about 5.2 kW to 25 kW in 5-kW increments, the thermal bus set point, as measured by a thermocouple on the two-phase return line, increased by about 2°F. Individual

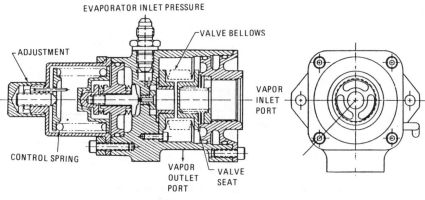

Fig. 16   Backpressure regulating valve (BPRV).

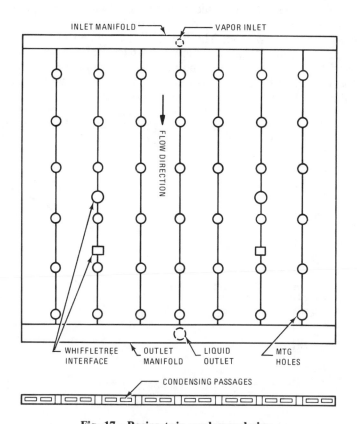

Fig. 17   Boeing twin condenser design.

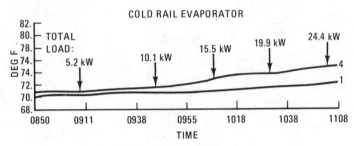

Fig. 18   Transient load effects (70°F): evaporator response.

evaporator outlet temperatures increased by 3–4°F. This is illustrated by Fig. 18. Similar behavior was exhibited at the 35°F set point.

System 10:1 Heat Load Turndown. A 10:1 heat load turndown (24.4–2.6 kW) was successfully executed at both the 35 and 70°F set points. Bus vapor temperature decreased by only 1°F during this period.

Quiescent Load Operation. The Boeing bus exhibited the ability to hold the set point within 2°F at a minimum heat load of 530 W at a 70°F set point and 1180 W at the 35°F set point. The higher minimum at the low set point reflects the difficulty of operating a bus at below ambient temperature, where ambient heat leaks become significant.

System Set-Point Temperature Control. The BPRV controls the bus set-point temperature by regulating the pressure at the liquid/vapor inter-

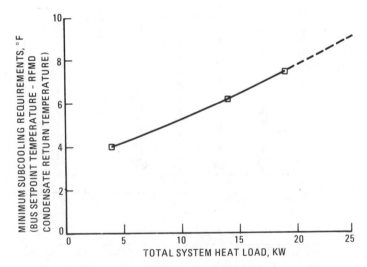

Fig. 19   Minimum subcooling requirements test at 70°F.

face inside the RFMD, thereby regulating the temperature of the evaporator supply liquid via the saturation pressure/temperature interdependence. During the course of the ambient test, the set point (temperature upstream of BPRV) was maintained to within 2°F of the desired set point. Individual evaporators were maintained within a somewhat larger isothermal band of approximately 4°F.

System Subcooling Requirements. Figure 19 presents a curve that shows the minimum subcooling requirements, defined as the minimum bus set point temperature minus the RFMD condensate return temperature for the bus under which set-point control was maintained. The curve shows the trend toward increased subcooling requirements at higher heat loads. Although the maximum load during this test was 19 kW, extrapolating the curve to the maximum system load of 25 kW yields a subcooling requirement of about 9°F. This trend will be similar for operation at other set points.

System Response to Variable Heat Sink Environments. Four simulated orbital cycles, each consisting of two orbits, were completed. Each cycle was conducted with a different heat load, the first with 25 kW, the second with 15 kW, the third with 5 kW, and the fourth with a variable load profile with loads ranging from 5 to 25 kW. Sink temperatures during these profiles varied from 15 to 45°F. Throughout this test, the system set point (BPRV vapor inlet temperature) was controlled to within 2°F.

Heat Load Sharing. The ability of an evaporator to load share, i.e., transfer heat to a payload or fluid heat exchanger in order to keep it warm, was demonstrated using the single-phase water heat exchanger. Heat (800 W) was transferred from the bus to the water loop, representing 15% of the total bus heat load.

System Pressure Drop. Maximum pressure drop in the two-phase return line was about 8 psid.

RFMD Electrical Power Usage. During the course of testing of the RFMD's power consumption ranged from 330 to 1000 W, with an average of 550 W of power being consumed.

System Transient Response During Set-Point Changes. The Boeing bus exhibited the ability to execute a set-point change from 70 to 35°F and also from 35 to 70°F. The time required to reach the new set point is dependent on the ambient environment, system heat load, and condenser sink temperature.

System Control Complexity. The BPRV provides system set-point control passively for operation at a given set point. A change of set point requires a change in the BPRV setting. Although this was accomplished manually during this test by means of a set screw, a flight system would have a remote capability to change the setting. Passive flow control to the evaporators was accomplished by the cavitating venturis. Liquid inventory control was provided passively by two metal bellows accumulators. A limitation of the liquid inventory control system is the response time in reducing the liquid level inside the RFMD during initial system startup and also after a large slug of liquid returns to the RFMD from the evaporator return line. The slow response in removing the liquid increases the RFMD drag and power consumption.

System NCG Sensitivity, Collection, and Venting Capability. During the set-point reduction from 70 to 35°F, NCG partial pressure in the bus increased, with the NCG being swept to and accumulating in the cold end of the RFMD. The NCG had to be vented a number of times to permit the set point to be reduced to 35°F. Following this initial venting, approximately 10.5 in.³ of NCG were injected upstream of the cold rail at set points of 35 and 70. This decreased the RFMD end-to-end pressure difference until the gas was vented. During this process, the system set point remained fairly stable.

Condenser Design Capacity. Figure 20 shows the amount of heat that can be rejected from both the shear-flow condensers and the twin Gregorig grooves condensers as a function of the average sink temperature for a bus set point of 70°F.

Evaporator and Heat Exchanger Performance. Surface temperature distributions indicated that the cold plate had an average surface-to-set-point temperature difference of 20°F at a 5-kW heat load, whereas the cold rails had a temperature difference of 20°F at a 2-kW heat load. This represents an effective heat-transfer coefficient of 130 Btu/h-ft²-°F for the cold plate and 192 Btu/h-ft²-°F for the cold rail. For the cold rail, the effective heat-transfer coefficient includes the mechanical contact resistance between a plate simulating equipment and the cold rail surface. In addition, both the cold plate and cold rails were successfully operated under conditions of nonuniform heat load.

The Boeing/Sundstrand thermal bus also underwent a thermal vacuum

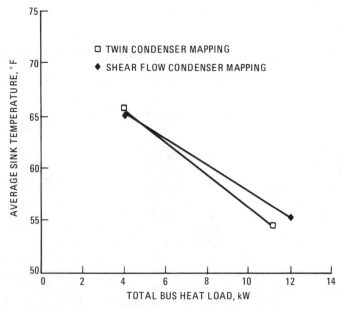

Fig. 20 Condenser mapping: sink temperature vs heat load (70°F bus set point).

test at NASA-JSC in July 1988. Shear flow condensers were coupled to three radiators provided by LTV, and the twin condensers were coupled to six radiators provided by Grumman (see the section on Central Heat Pipe Radiator System for a discussion of the different radiator designs). The bus demonstrated its ability to operate with a radiation heat sink over a range of operating conditions, although problems with the LTV radiators limited some of the test points that could be accomplished.

One phenomenon that was evident in the T/V test was the influence of natural circulation in the twin condenser manifold. This caused the condensers to be loaded sequentially, rather than in parallel. At low loads, cold outlet fluid from the on-line condenser would flow back through the other condensers in reverse flow. At very low loads, this reverse flow would join the vapor going into the on-line condensers and condense it, so that the inlet temperature dropped below saturation. As the load was increased on the system, the pressure drop through the on-line condensers increases as the liquid/vapor interface moved toward the exit. When this driving potential was high enough, the natural circulation would be overpowered and another twin condenser would come on-line. The potential problem caused by this behavior is that the radiators will be subjected to a sudden temperature spike as each condenser comes on-line and warm vapor replaces the cold fluid. Depending on the radiator design, this could result in a full or partial dryout from which the radiator may or may not recover.

In addition to ground testing, a small thermal bus loop built by Sundstrand under contract to NASA-JSC was flown on the NASA-JSC KC-135 reduced-gravity aircraft in April 1987 for the purpose of observing and measuring two-phase flow in reduced gravity. The loop, which used Freon-114 as the working fluid, consisted of an RFMD, a BPRV, a bellows accumulator, a swirl flow evaporator, a shear-flow condenser, and a clear section on the two-phase return line, is shown in Fig. 21. Test results and conclusions regarding two-phase flow pressure drop in both 1 g and 0 g, as well as analytical modeling techniques and their correlation with the data, are discussed in Ref. 1 and are summarized in the following list:

1) For two-phase flow inside tubes, the pressure drop is strongly affected by the flow regime.

2) Pressure drops occurring in two-phase flow under reduced gravity conditions are significantly greater (by up to a factor of three) than those under 1-g conditions for the same conditions of mass flux and quality.

3) At vapor qualities greater than about 15%, the observed flow pattern was annular. Taylor bubbles were observed at qualities less than about 15%.

Grumman Mixed-Flow Thermal Bus (GMFTB) Concept

A second concept for a mixed two-phase flow thermal bus has been proposed by Grumman and is shown schematically in Fig. 22. The GMFTB utilizes the same basic hardware and set-point control method as the Grumman separated bus but replaces the reservoirs, evaporator liquid supply valves, and ultrasonic liquid presence sensors with a fixed orifice upstream of each evaporator. The orifice maintains an approximately

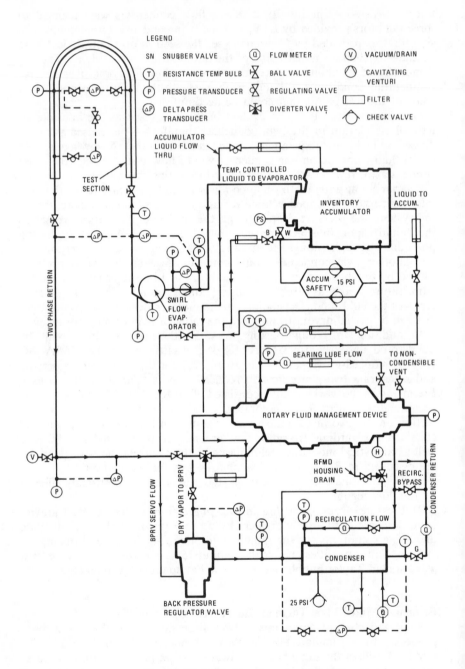

Fig. 21   KC-135 test loop schematic.

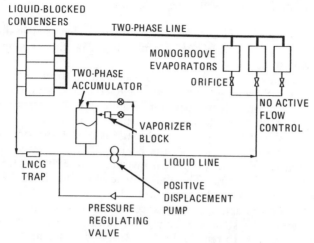

Fig. 22   Grumman mixed two-phase thermal bus.

constant flow to each evaporator. Liquid is vaporized from the circumferential wall grooves of the monogroove evaporators, which can be identical to the evaporators used in the Grumman separated thermal bus (see Fig. 2). At lower heat loads, the evaporators will be partially flooded; thus, the evaporator heat-transfer coefficients may be lower than in the separated system. However, at high loads, which are the most critical case in terms of heat-transfer temperature difference, the heat-transfer coefficients will approach those of a separated system.

Depending on the heat load, a two-phase mixture of varying quality will exit the evaporator. This mixture flows in a two-phase return line to the condensers, described previously and shown in Fig. 5. Liquid blockage, controlled by the two-phase accumulator, maintains the system operating temperature. For the GMFTB, the two-phase accumulator tank must be larger to accommodate the liquid inventory needed to partially fill the two-phase return line under conditions of low system heat loads.

Identical to the Grumman separated bus, a constant-speed, positive displacement pump transports the liquid to the evaporators, with excess flow circulating in a bypass loop. Pump outlet pressure is kept constant by a passive, spring-loaded relief valve. In order to reheat the liquid entering the evaporators to near the saturation temperature, a regenerator may be required. This would be a tube in tube heat exchanger that transfers heat between warm vapor in the two-phase return line and the subcooled liquid leaving the pump. Reheating of the subcooled liquid may be necessary so that evaporation, rather than just sensible heating, always occurs in the evaporators, even at very low heat loads.

Grumman Mixed-Flow Thermal Bus Test Data

Using the hardware developed for the GPTBS, the GMFTB underwent limited testing at the conclusion of the GPTBS ambient test at

NASA-JSC in January 1989. The following test points were successfully achieved.

System Transient Load Effects. Heat load was increased in about 5-kW increments from 5 to 26 kW. As shown in Fig. 23a, the set point was held constant within 1.5°F.

System Turndown of 10:1. Heat load was reduced from 22.5 to 2.5 kW in a step change. System set point was held constant, and there were no noticeable effects on system operation. Figure 23b illustrates the response of the vapor temperature to the load turndown of 10:1.

System Load Biasing. Various combinations of evaporators were loaded to determine if the placement of the heat loads impacted system performance. No adverse effects on system operation were detected.

Evaporator Load Biasing. The ability to accept nonuniform heat loads of up to 5 kW was successfully demonstrated.

Condenser Performance. Performance of the condensers in the mixed system as well as system subcooling requirements were evaluated. The condensers were able to reject a load of 23 kW with a sink temperature of 50°F. As expected, these results are similar to those obtained with the Grumman separated bus. System subcooling requirements were determined with and without the regenerator operating. At a load of 22.8 kW, without a regenerator, the system operated with only 3°F of subcooling. Utilizing the regenerator increased the subcooling requirement to 9–15°F at similar heat loads.

Rapid Heat Load Transients. The bus easily handled rapidly varying evaporator heat loads. Step changes in evaporator heat loads up to 5.25 kW were made, with the total system load varied from 5 to 21 kW. Set-point control was maintained throughout all transients.

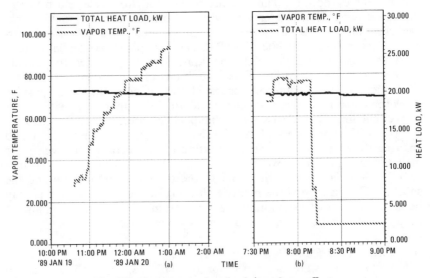

Fig. 23   System transient loads/turndown effects.

70 – 35°F Set-Point Reconfiguration. A successful set-point reconfiguration from 70 to 35°F was demonstrated.

Pump Power Consumption. Pump power consumption was approximately 37 W.

Comparison of Thermal Bus Concepts

Both the Boeing/Sundstrand and Grumman separated-flow buses have undergone extensive testing and have shown the ability to meet all of the Space Station requirements. The Grumman mixed bus, although not as extensively tested, has also shown promise that it too will meet all of the Space Station requirements. The Lockheed bus has not yet been tested; thus, no conclusion can be drawn regarding its ability to meet Space Station requirements. Table 1 compares the features of the various systems.

Each system has certain strengths and weaknesses. Whereas the Boeing/Sundstrand bus eliminates the need for evaporator flow control and capillary fluid transport, the RFMD and BPRV are complex components that must be shown to meet the long-life and high-reliability requirements of the Space Station. In addition, the RFMD power consumption is high compared to the other systems. Although little 0-*g* two-phase flow data are available, the mixed two-phase flow nature of the system will yield significantly higher pressure drops and require a larger fluid inventory than would a separated system.

The Lockheed bus, although it requires less pumping power than the Boeing bus, utilizes very fine (1-μ pore size) wicks, resulting in an increased susceptibility to contamination. Although only a few valves are required for control, these valves are complex modulating valves. The software required to maintain system control needs to be demonstrated. Although phase separation at the evaporators is designed to be maintained by passive rather than active means, the ability of the wicks to accomplish the separation over the full range of heat loads needs to be demonstrated. In addition, this method requires that a third transport line for recirculation of excess liquid be routed throughout the Space Station. Finally, the use of gas blocked condensers, although allowing 1-*g* test data to be more accurately extrapolated to 0 *g*, risks contamination of the rest of the bus with noncondensable gas.

Both the Grumman separated- and mixed-flow concepts require very low pumping power. The separated system avoids two-phase flow issues, but at the expense of providing a reservoir, ultrasonic sensor, and on/off valve in conjunction with each evaporator. The mixed-flow system eliminates these but carries the two-phase flow penalties discussed previously. Use of the two-phase accumulator simplifies the accumulator design by obviating the need for a metal bellows but will require some internal baffling to maintain proper fluid positioning in 0 *g*. Additionally, the Grumman concept provides design flexibility since the same basic hardware can be used in either a mixed- or a separated-flow mode.

Finally, the applicability of 1-*g* test data to 0-*g* operation should be addressed, both in terms of the various systems and their individual components. The behavior of two-phase flow in 0 *g* represents the greatest

Table 1 Comparison of thermal bus features

Features	Lockheed	Boeing/ Sundstrand	Grumman separated flow	Grumman mixed flow
Type of flow	Separated	Mixed	Separated	Mixed
Pumping device/power consumption (25 kW bus)	Centrifugal/ 75–200 W	RFMD/ 500–1000 W	Positive displacement/ 40 W	Positive displacement/ 40 W
Evaporator flow control	Passive: orifice/wicks	Passive: cavitating Venturi	Active: ultrasonic sensor/valve	Passive: orifice
Set-point control	Modulating valves/ gas-blockage	BPRV	Liquid blockage	Liquid blockage
Type of accumulator	Metal bellows	Metal bellows	Two-phase	Two-phase
No. of fluid transport lines	3	2	2	2
Strong capillary dependence?	yes	no	yes	no

uncertainty for the mixed-flow systems. Unless numerous 0-g flight experiments are carried out to evaluate the effects of bends, fittings, etc., system sizing will have to be very conservative. For the Boeing/Sundstrand thermal bus, 0-g component behavior should be similar to 1-g behavior due to the forced-flow nature of the system, although interactions of the condensers and the radiators in 0 g remains an issue. The Grumman mixed-flow bus shares the two-phase flow issues and in addition requires KC-135 reduced-gravity testing to verify proper 0-g fluid positioning in the two-phase accumulator.

Both the Grumman and Lockheed separated-flow buses rely on capillary forces in the evaporators. Although the effects of gravity can be minimized during ground testing, providing some degree of confidence in 0-g operation, ground tests do not eliminate the need for 0-g verification. The Grumman concept requires adequate demonstration of proper fluid management within the reservoir, as well as transport from the reservoir to the monogroove evaporators. The Lockheed bus requires verification of the ability of the evaporators to properly distribute the flow between the evaporation sites and the liquid recirculation lines as well as to provide phase separation.

Ground tests of the Grumman liquid-blocked condenser have indicated that the fluid distribution should be similar in 0 g to 1 g. An integrated ground test with radiators would therefore provide a high degree of confidence that the bus condenser and radiator will operate properly in 0 g, although a 0-g verification would be desirable, since this is a critical interface.

For the Lockheed bus, the 1-g gas-blocked condenser performance would be more closely related to 0-g performance than would the liquid-blocked condenser performance. In either case, condenser/radiator interaction in 0 g remains an issue.

Central Heat Pipe Radiator System

General Design Description

The central heat pipe radiator system, also known as the SERS, is being developed by NASA to provide a reliable, maintainable, long-life waste heat rejection capability for the Space Station and similar large space systems. In general, the SERS features modular, high-capacity heat pipe radiator panels that can be installed and replaced on orbit via remote manipulator or extravehicular activity (EVA). Each panel interfaces with the central heat-transport loop through a dry contact heat-exchanger attachment. The contact heat-exchanger assembly includes the thermal bus condenser and a clamping mechanism, which supplies pressure at the dry contact interface with the evaporator section of the radiator panel. The basic concept is shown in Fig. 24, and prototype designs are under development by two NASA contractors, Grumman and Lockheed/LTV.

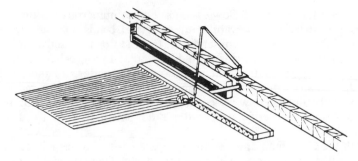

SERS PANEL INSTALLATION WITH MOBILE REMOTE MANIPULATOR SYSTEM

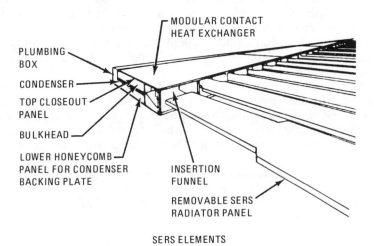

SERS ELEMENTS

Fig. 24 SERS concept.

The Space Station system requirements that drive the design of both SERS concepts are listed below:

1) Total system heat load = 70 kW initial operating configuration (IOC) = 20 kW at $2 \pm 2.5°C$ and 50 kW at $21 \pm 10°C$;

2) Heat load range: MAX/MIN = 10/1;

3) 10-yr system life with 99% reliability;

4) Minimum heat-rejection area (radiator fin effectiveness > 90%); and

5) Economical size compatible with Orbiter cargo bay packaging.

Grumman SERS

Radiator Panel

The prototype SERS being developed by Grumman is based on a low-risk extension of proven monogroove heat pipe technology. The overall geometry was determined from previous system trade studies[14] and practical considerations such as ground handling and test chamber restrictions. Each of the 1-ft-wide SERS radiator panels consists of a compact six-leg evaporator section (2 ft long) that is joined to a dual-leg, 46-ft-long condenser section by welding to a common manifold (Fig. 25). The condenser/radiator fin is a simple sheet and stringer design consisting of the two heat pipe legs and edge channels sandwiched between two aluminum facesheets (Fig. 26). Conductive epoxy adhesive is used to bond the structural sandwich together. The double-sided radiator has been designed for a minimum fin effectiveness of 90% in order to minimize the required radiator area for an operational space station system. Each panel also includes a unique liquid trap section at the end of the condenser which is designed to retain all excess liquid within the heat pipe to prevent it from accumulating and blocking the active condenser section. This feature permits a fully operational heat pipe condenser over an extreme range of temperatures, from -90 to $+70°F$. The alternative solution to this potential loss of radiator area would be to include extra radiator elements that would result in a greater weight penalty.

The key to the Grumman/NASA SERS panel is the modular construction concept wherein the radiator panel is made up of separate subassemblies that are designed for manufacturing and handling ease. There are six primary subassemblies for each panel: the multileg evaporator section, four identical 10-ft-long condenser/fin sections, and the closure condenser section that contains the liquid trap.

The evaporator subassembly consists of six individual heat pipe legs, each containing a screen mesh insert, that are welded to a common distribution header. The mesh insert, used only in the evaporator, serves two purposes: preservation of liquid channel subcooling at high heat fluxes

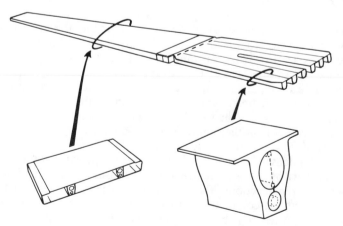

Fig. 25   Space erectable radiator panel: GTA hardware.

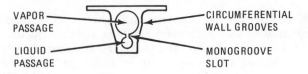

VAPOR PASSAGE
LIQUID PASSAGE
CIRCUMFERENTIAL WALL GROOVES
MONOGROOVE SLOT

HEAT PIPE DETAIL (FLANGE REMOVED IN CONDENSER)

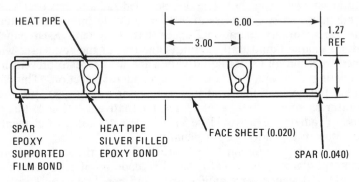

HEAT PIPE
6.00
3.00
1.27 REF

SPAR EPOXY SUPPORTED FILM BOND
HEAT PIPE SILVER FILLED EPOXY BOND
FACE SHEET (0.020)
SPAR (0.040)

Fig. 26   Grumman SERS: condenser/radiator fin details.

and augmentation of the liquid flow between the monogroove slot and the circumferential wall grooves within the evaporator vapor channel. The ends of the heat pipe tubes are sealed by aluminum plugs that are welded into each channel.

Each evaporator leg is designed to be as structurally independent as possible, requiring only small stitch welds at the extremes of the adjacent flanges. This increases flexibility in the long direction and allows better interface contact by permitting each leg to deflect independently as local conditions dictate.

The flange surfaces of the evaporator section are prepared by machining the base material, flame spraying with pure aluminum, and then grinding the sprayed surface to the desired thickness. Surface finish and flatness specifications are 16–32 in. rms and 0.001 in./in., respectively. The suitability of the contact interface of each evaporator subassembly is checked by using a duplicate clamp test fixture and special pressure sensitive film to record the image at the clamping interface under the 100-psi design load. These image results are then compared with precalibrated references obtained from component thermal tests that relate the image to corresponding measured values of contact conductance.

In each condenser/fin subassembly, the heat pipe extrusion is treated as an integral part of the section structure. It consists of a bottom facesheet, the two heat pipe sections, the edge closure channels, and the top facesheet. A silver-filled conductive epoxy bond is used on the heat pipes and an epoxy supported film bond on the channels. Both cure at a nominal temperature of 325°F under moderate pressure.

The liquid trap section consists of a dead-ended length of heat pipe extrusion connected to the ends of the active condenser sections. During 0-*g* operation, any excess liquid (due to density variations over the operating temperature range) will accumulate and be retained within the trap vapor channel. This excess is free to re-enter the liquid channel through the normal capillary paths whenever needed.

Only after final assembly, when all of the heat pipe sections are butt welded together, does the panel contain a true hermetically sealed heat pipe. The structural connection between adjacent edge channels and face sheets is made with simple flush-riveted splice plates. Thus, the final processing and acceptance testing of the heat pipe is done only once and occurs as an integral part of the final checkout of the completed radiator assembly.

Design data for the SERS radiator panels under development for the Space Station are summarized in Table 2. Performance characteristics are summarized in Fig. 27.

Heat Pipe

The monogroove concept for an improved high-performance heat pipe design permits high heat-transport capacity without impacting heat-transfer efficiency. It combines the advantages of axial grooves, such as simple construction and large liquid and vapor areas, with the high heat-transfer

Table 2   Design data summary for Grumman SERS radiator panels

	Total	Evaporator	Condenser
• Geometry			
— Length (ft)	47.9	2.1	45.8
— Width (in.)	—	12	12
— Thickness (in.)	—	1.22	1.27
• Weight (lb)	116.7	9.7	107
• Heat pipe (hp) concept:	Grumman monogroove		
• Construction details			
— Evaporator:	— 6 independent legs welded to common header		
	— Contact flange clad with soft aluminum/16 microinch finish		
— Condenser:	— 2 parallel legs with separate excess liquid trap section		
— Radiator fin:	— Simple bonded sheet-stringer		
	— 10 ft subassemblies spliced together		
• Materials			
— Hp:	Aluminum 6061-T6/ammonia		
— Sheetstock:	Aluminum 6061-T6		
— Adhesives:	EA9601/Reliabond 350		
— Thermal coating:	Chemglaze A276 (ground)		
	Silver teflon (flight)		

- TEMP RANGE: -90° TO 140°F

- ACCEPTANCE TEST HEAT TRANSPORT: 2.4 KW @ 1/2 IN. TILT

- OVERALL THERMAL CONDUCTANCE: 5.0 $\frac{°F}{KW}$ (② TO ③)

  7.0 $\frac{°F}{KW}$ (① TO ③)

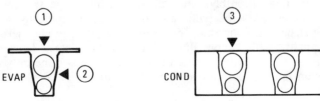

- η = 0.93 (FOR T_{ROOT} = 70° F, T_S = -100°F)

- WEIGHTS: EVAPORATOR   = 8.7 LB
             CONDENSER   = 103.2 LB (HP ~ 50%, FIN ~ 27%)
             AMMONIA FLUID = 4.8 LB
             TOTAL   = 116.7 LB (~ 2.4 LB/SQ FT)

- MAX EVAPORATOR HEAT FLUX = 42 W/IN^2 OF FLANGE AREA

Fig. 27   Performance characteristics SERS: 2-kW baseline radiator panel.

coefficients of circumferential wall grooves. The monogroove design (Fig. 28) contains two large axial channels, one for vapor and one for liquid. The small slot separating the channels creates a high capillary pressure difference, which, coupled with the minimized flow resistance of the two separate channels, results in the high axial heat-transfer capacity. The high evaporation and condensation film coefficients are provided separately by circumferential grooves in the walls of the vapor channel without interfering with the overall heat-transport capability of the axial grooves.

Radiator panels of various configurations have been made for the past 5 yr using the "baseline" monogroove heat pipe extrusion show in Fig. 28 that has a nominal 25,000-W-m transport capacity. Two 40-ft-long, U-shaped panels were successfully thermal vacuum tested (2.4-kW maximum load) at NASA-JSC in February 1984[6,15] and continue to serve as life test articles with performance verification run on scheduled 6-m intervals. A special monogroove heat pipe radiator experiment was flown on Shuttle flight STS-8 (August 1983) and demonstrated successful startup and sustained operation in the Shuttle payload bay environment.[16,17] Most recently, three full-size (50-ft-long) monogroove radiator panels were built and certified for the space station heat pipe advanced radiator element (SHARE) Shuttle flight experiment. All of the flight certification tests were achieved, including a maximum heat rejection of 2 kW, thawing of a frozen panel, and continuous operation under cycling environmental and evaporator heat loads.[18] The SHARE flight experiment is flown on STS-29 (March 13, 1989). Evaluation of the flight data uncovered problems with manifold priming and

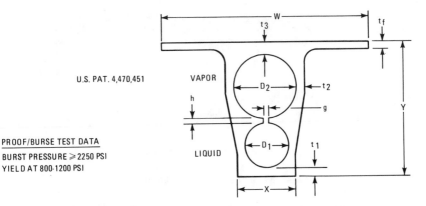

U.S. PAT. 4,470,451

PROOF/BURSE TEST DATA

BURST PRESSURE ⩾ 2250 PSI
YIELD AT 800-1200 PSI

MATERIAL: 6061-T6
DRY WT: 0.61 LB/FT (W/FLANGE), 0.51 LB/FT (W/O FLANGE)
V_L = 24.0 CC/FT
V_V = 53.2 CC/FT

	DIMENSIONS, IN.									
	D_1	D_2	g	h	t_1	t_2	t_3	t_f	W	X	Y
AS SPECIFIED	0.398	0.586	0.008	0.052	0.077	0.091	0.108	0.065	1.970	0.558	1.221
(TOLERANCE)	±0.002	±0.002	±0.003	±0.002	±0.002	±0.002	REF	±0.005	±0.005	±0.016	±0.010
SAMPLE MEASUREMENT	0.388	0.580	0.010	0.047	0.086	0.094	0.112	0.068	1.966	0.559	1.218

THERMAL PERFORMANCE TEST DATA

● EVAPORATOR HEAT FLUX (W/IN^2) = 42

● HEAT TRANSPORT CAPACITY (KW-IN)
  – SINGLE LEG CONDENSER = 660.
  – DOUBLE LEG CONDENSER = 840.

Fig. 28   Details of baseline monogroove heat pipe extrusion.

bubble management that could only be revealed in extended 0-*g* testing of a full-size unit. Design modifications have been developed to ensure the same performance under both 1-*g* and 0-*g* conditions. These changes are now being validated in a series of 0-*g* experiments for the KC-135 aircraft and the shuttle mid-deck. Another full-size radiator test article, SHARE II scheduled for January 1991, will provide final confirmation.

Clamping Mechanism
   The required uniform contact pressure at the dry thermal interface between the heat pipe evaporator section and the thermal bus condenser heat exchanger is provided by a simple lightweight clamping mechanism known as a whiffletree. It features a network of precisely spaced, interconnected, aluminum beam elements that takes a single centrally applied load and distributes it equally among a number of predetermined pressure points. In

a) Manual drive unit—top view

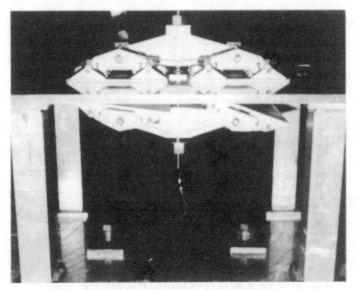

b) Manual drive unit prior to panel insertion—side view

Fig. 29   Whiffletree clamp.

the SERS application, the whiffletree clamp (Fig. 29) is designed to produce a contact pressure of 100 psi over a 2-ft^2 area by using a network of 48 pressure points, which corresponds to 8 points along each of the six legs (spaced every 3 in.). The pressure is applied to the narrow side of the heat pipe section and presses the opposite flanged surface against the facesheet of the mating heat exchanger. Each leg acts like a beam between the points of concentrated load, so that the contact along the flange is relatively uniform.

The central load is applied by a high-strength steel bolt that can be tensioned by operating a motor-driven manual nut on either side of the unit, for redundancy. Operating characteristics and performance data are summarized in Table 3.

Table 3 Characteristics of SERS wiffletree clamp

Description	— Simple network of connected beam elements
	— 48 pressure points (8 per hp leg)
	— Single centrally located high strength steel bolt (4340 or mp 35)
	— All-mechanical system
	— Redundant operation
	— No power or external forces required after clamping
	— Only two penetrations of bus condenser needed (bolt, translation guide)
Performance data	
— Contact area	: 1 ft × 2 ft
— Volume envelope	: 22 in. l × 11 in. w × 17 in. h
— Contact pressure	: 100 psi
— Clamp load	: 28,800 lb/2000 in.-lb
— Clamp time w/ motorized tool	: 3 to 4 min
— Weight	: 30 lb
— Translating stroke	: 2.75 in.

Lockheed/LTV SERS

Radiator Panel

Figure 30 shows the Lockheed/LTV SERS radiator panel design. Each panel is 12 in. wide, approximately 1.1 in. thick, and 48.4 ft long. The contact heat-exchanger end of the panel is tapered to 9 in. wide to facilitate insertion of the panel into the contact heat-exchanger module. Two high-capacity heat pipes are contained in each panel. Each heat pipe has three evaporator legs at the 9-in.-wide end of the panel (6 evaporator legs per panel) and a liquid reservoir at the condenser end. The liquid reservoir allows for thermal expansion or contraction of the heat pipe liquid and is sized to allow heat pipe operation with temperatures as low as $-80°F$. Heat is absorbed in the evaporator legs from the thermal transport loop condenser through the contact heat-exchanger surfaces on the radiator panel and the thermal transport loop condenser. The heat pipe transfers the heat out to the radiator surface where it is radiated to space. Two heat pipe condenser legs are contained in each panel.

An adhesively bonded facesheet and honeycomb construction is used to provide a weight-efficient, high-radiation-effectiveness fin with good stiffness and handling characteristics.

The honeycomb acts as both a structural stiffener and a thermal conductor. Tests conducted during the Orbiter radiator development confirmed that the lateral conduction in the honeycomb enhances the radiation fin effectiveness. The honeycomb also provides a conduction path for facesheet-to-facesheet conduction to reduce temperature differences due to

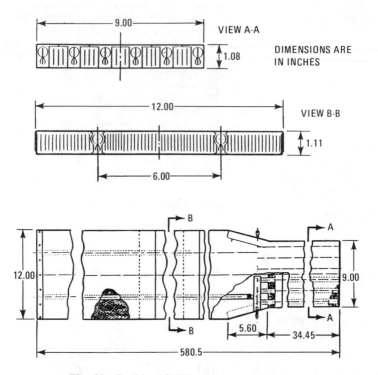

Fig. 30 Lockheed/LTV radiator panel design.

different environments on the radiator faces and the different thermal resistances from the heat pipe condenser to the facesheets.

A fiberglass honeycomb is used between the heat pipe evaporator legs to support the crush loads of the contact heat exchanger. The fiberglass honeycomb reduces the heat transfer to the liquid artery in the evaporator section of the heat pipe and helps to maintain the high evaporator heat flux capability.

Characteristics of the SERS radiator panel being developed for ground test bed validation are contained in Table 4.

Heat Pipe

The SERS heat pipe is an extended design of the Lockheed tapered artery heat pipe, which has undergone continuous development since 1981. The heat pipe (Fig. 31) has a composite wick with a liquid artery located adjacent to a vapor channel. Circumferential wall grooves provide liquid communication between the liquid artery and the heat-transfer surface of the vapor channel.

The final design configuration is shown in Fig. 32. The 45-ft single condenser leg is fitted with a reservoir through a machined adapter that

Table 4   Characteristics of SERS radiator panel

Length (total panel)	48 ft
Condenser width	12 in.
Evaporator configuration	Flat contact interface
Evaporator width	9 in.
Evaporator length	27.5 in.
Condenser length	44.74 ft
Panel thickness	1.10 in.
Heat pipe	Lockheed tapered artery
Heat pipe configuration	6 evaporator legs & 2 condenser legs
Panel weight	94 lb
Heat rej per panel at max environment	1.5 kW
Heat pipe evaporator flux at worst case	11 Watts per in.
Radiation fin effectiveness	0.9
Radiator fin construction	Aluminum honeycomb

contains one of two fill tubes on the pipe. The reservoir is sized to store all excess liquid resulting from fluid density changes over the required operating temperature range.

Three parallel evaporator legs are attached to the condenser through a machined manifold that contains circular passages for each fluid. No vapor/liquid interface is provided within the manifold. The liquid passage path from condenser to evaporator is carefully designed to ensure artery priming under microgravity conditions. A second fill tube is located on the evaporator manifold.

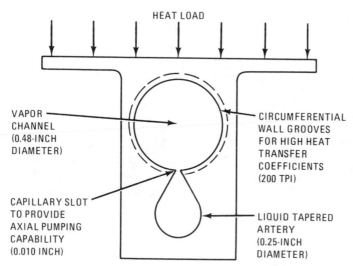

Fig. 31   Lockheed tapered artery heat pipe cross section.

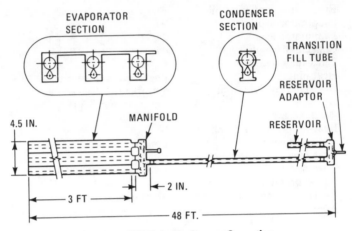

Fig. 32   SERS heat pipe configuration.

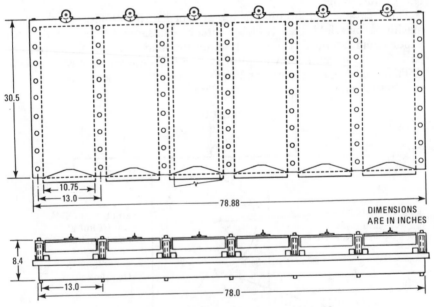

Fig. 33   SERS contact heat-exchanger assembly.

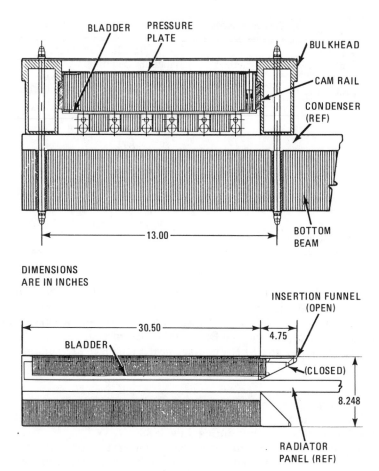

Fig. 34   Contact heat-exchanger cross sections.

Clamping Mechanism
An overall view of the SERS prototype contact heat exchanger is illustrated in Fig. 33. Figure 34 shows the module cross sections. The radiator panels are inserted into the contact heat exchanger during the on-orbit buildup. An insertion funnel is used to guide the panel into the heat-exchanger opening. After the panel is inserted, an actuation cam in the heat exchanger is turned 180 deg to move the pressure plate backward against the bulkhead cam rails. This, in turn, moves the pressure plate and pressurizing bladder down approximately 1 in. to snug the radiator panel against the condenser surface. The bladder is pressurized with nitrogen to 130 psi to provide the contact force between the radiator panel and condenser. The bladder configuration is a welded assembly fabricated from two flat sheets of 0.050-in. titanium. Each sheet is chem-milled to 0.01 in. except around the edge where the thickness remains at 0.05 in. Attached to the bladder are six

Table 5   Characteristics of SERS contact heat exchanger

Description	Flat contact heat exchanger
Contact pressurizing method	Titanium diaphragm pressurized with nitrogen gas pressure
Modularity	One condenser unit interface multiple radiator panels (six radiators baselined per condenser module and two per subcooler)
Contact area	9×27.5 in.
Approx. interface weight	42 lb per panel
Contact pressure per panel	130 psi beginning of life
Projected contact conductance	925 Btu/hr-ft$^{2\circ}$-F beginning of life 740 Btu/hr-ft$^{2\circ}$-F end of life (10 yr)
Contact temperature drop	4.8°F for end of life

attachment bosses and a fill tube. The radiator side of the bladder is covered by a 0.005-in. stainless steel scuff plate. Component test data have confirmed that this clamping system can achieve an interface contact conductance exceeding the 500 Btu/h-ft$^{2\circ}$-F design goal.

Characteristics of the SERS contact heat exchanger are summarized in Table 5.

NASA Thermal Test Bed Results

Both candidate SERS configurations were thermal vacuum tested in the NASA-JSC test bed facility during July 1988.

Two types of tests were run: one was an integrated test with the Boeing prototype two-phase thermal bus, and the other was a stand-alone evaluation using electrical heat input to the evaporator section. The integrated test evaluated the overall performance of a prototype space station active thermal-control system, including two-phase heat transport and acquisition, high-capacity heat pipe radiator panels, and the dry attachment interface provided by the clamping mechanism. Of special interest was the test to simulate the effects of attaching a cold "replacement" panel to an operating thermal bus condenser by selectively loosening and reclamping the units under in situ thermal/vacuum conditions.

The stand-alone test of a single panel was run to correlate performance to accurately measured test conditions since the integrated test setup lacked the necessary instrumentation to determine the individual heat load to each panel.

The independent SERS panel test ensured a more accurate knowledge of the following critical parameters: evaporator heat input (power measurement at the electrical heater); adverse tilt (direct unobstructed measurements at selected points on the test panel during testing by use of a surveyor's transit); and environmental heat flux (radiometer measurements in a cold, uncluttered environment). The test points consisted of measuring maximum heat-transfer capability as a function of adverse tilt and the corresponding overall thermal conductance between evaporator and fin root. Transient response as a function of varying evaporator and environmental heat loads was also measured.

The steady-state performance map for the Grumman stand-alone test panel is shown in Fig. 35. Heat transfer in excess of 3 kW at 0.43 in. adverse tilt was achieved. Of greater interest is the comparative behavior of the two test panels in response to stepped load changes. This is shown in Fig. 36 for tests run over a nominal 8-day period. Under the same test conditions, there were five instances where the Lockheed/LTV test panel deprimed and failed to operate. The Grumman test panel functioned properly in all instances.

Repriming of the failed Lockheed/LTV panel was achieved by first removing the heat load, then increasing the environmental heat flux until the condenser temperature exceeded that of the evaporator. This reversed the normal direction of the heat pipe cycle and returned liquid to the dried out evaporator section via the condensation process.

The fin effectiveness values calculated by NASA for the two panels were 0.94 for the Grumman design and 0.88 for Lockheed/LTV.

The test results for the integrated testing are still being analyzed by NASA and the contractors. Although the integrated thermal bus/SERS operated successfully at loads up to 14 kW, direct comparisons are difficult since

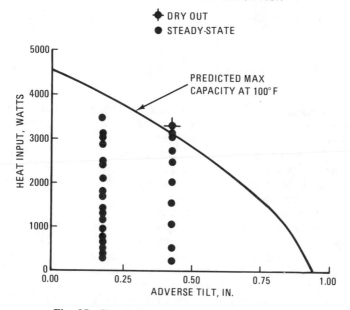

Fig. 35 Stand-alone thermal vacuum test data.

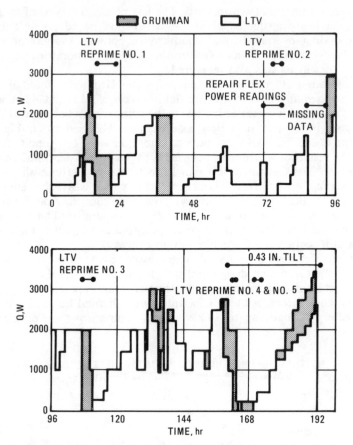

Fig. 36   Stand-alone power profile first half of T/V test 0.18 in. adverse tilt.

different condenser configurations were used with each SERS design. The six Grumman SERS panels were attached to separate Gregorig groove condensers, whereas the three Lockheed/LTV SERS panels were attached to a single shear-flow condenser. Apparently flow balancing problems within the two-phase bus caused unequal load distribution and the inability to conveniently test with both sets of panels operating together. Most of the integrated testing was done separately.

Transient response of a Grumman SERS panel is illustrated in Fig. 37 for the test series that simulated the loss and subsequent reattachment of a single radiator panel. Release of the panel causes a brief perturbation in bus set point. The panel itself cools uniformly to the sink temperature.

Upon reattachment, the evaporator section is initially thermally shocked but quickly recovers proper heat pipe function as both evaporator and condenser sections increase in temperature uniformly to their original condition.

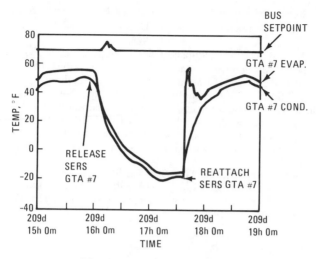

Fig. 37   Loss of SERS panel.

References

[1]Chen, I., Downing, R., Keshock, E., and Al-Sharif, M., "An Experimental Study and Prediction of a Two-Phase Pressure Drop in Microgravity," AIAA Paper 89-0074, Jan. 1989.

[2]"A Study of Two-Phase Flow in a Reduced Gravity Environment Final Report," Sundstrand Energy Systems, Oct. 16, 1987.

[3]Edelstein, F. and Liandris, M., "Thermal Test Results of the Two-Phase Thermal Bus Technology Demonstration Loop," AIAA Paper 87-1627, June 1987.

[4]"Grumman Prototype Thermal Bus Ambient Test Requirements Document," NASA-JSC Crew and Thermal Systems Division Document CTSD-SS-199, Aug. 24, 1988.

[5]Brown, R., Gustafson, E., and Parish, R., "Design of an Ammonia Two-Phase Prototype Thermal Bus for Space Station," Society of Automotive Engineers Paper 871506, July 1987.

[6]Alario, J., Brown, R. and Otterstedt, P., "Space Constructible Radiator Prototype Test Program," AIAA Paper 84-1793, June 1984.

[7]"Prototype Space-Station Space-Erectable Radiator System (SERS) Ground Test Article Development Final Report," Grumman Space Systems Division, Nov. 1988.

[8]"Grumman Prototype Thermal Bus Ambient Test Quick-Look Report," NASA-JSC Crew and Thermal Systems Division Document CTSD-SS-TBD, Jan. 27, 1989.

[9]"Lockheed Missiles and Space Company Ground Test Unit Prototype Thermal Bus Ambient Test Requirements Document," NASA-JSC Crew and Thermal Systems Division Document CTSD-SS-281, Dec. 2, 1988.

[10]"Lockheed Thermal Bus Ambient Test Final Report," NASA-JSC Crew and Thermal Systems Division Document CTSD-SS-183, Nov. 9, 1987.

[11]Myron, D. and Parish, R., "Development of a Prototype Two-Phase Thermal Bus System for Space Station," AIAA Paper 87-1628, June 1987.

[12]"Boeing Prototype Thermal Bus Ambient Test Final Test Report," NASA-JSC Crew and Thermal Systems Division Document CTSD-SS-284, Nov. 3, 1988.

[13]"Space Station Prototype Two-Phase Thermal Bus System (TBS) Final System Evaluation Report," Boeing Aerospace Document D180-31234-1, Engineering Technology, Boeing Aerospace, Sept. 30, 1988.

[14]Brown, R. and Alario, J., "Space Constructible Radiator System Optimization," Society for Automotive Engineers Paper 851324, July 1985.

[15]Marshall, P. F., "Space Constructible Heat Pipe Radiator Thermal Vacuum Test Program," Society for Automotive Engineers Paper 840973, July 1984.

[16]Rankin, J. G., "Integration and Flight Demonstration of a High Capacity Monogroove Heat Pipe Radiator," AIAA Paper 84-1716, June 1984.

[17]Alario, J., "Monogroove Heat Pipe Radiator Shuttle Flight Experiment: Design, Analysis, and Testing," Society for Automotive Engineers Paper 840950, July 1984.

[18]Alario, J. and Otterstedt, P., "The SHARE Flight Experiment—An Advanced Heat Pipe Radiator for Space Station," AIAA Paper 86-1297, June 1986.

[19]Oren, J. and Holmes, H., "Space Erectable Radiator System Development," AIAA Paper 88-0469, Jan. 1988.

Capillary Pumped Loop Supporting Technology

C. E. Braun*

General Electric Company, Princeton, New Jersey

Development of the enabling technology for space platform thermal control has proceeded via a series of well-documented tests[1,2] and will culminate in a Shuttle flight test of a prototype capillary pumped heat acquisition/transport/rejection loop (CAPL), developed by General Electric (GE) Astro-Space Division and OAO Corporation under funding from NASA/GSFC.[3] Complementing the hardware development are the synthesis of computer software tools to characterize two-phase fluid loop behavior in ground test and microgravity conditions. The status of both hardware and software development is discussed later in the paper.

Hardware

A generic capillary pumped loop (CPL) consists of a wicked evaporator and a continuous loop that allows cocurrent flow of the vapor and liquid (Fig. 1). As the evaporator is heated, the liquid evaporates from the saturated wick and flows as a vapor through the loop to the condenser zone, where heat is removed. The subcooled fluid is then returned to the evaporator.

The CAPL flight experiment schematic is shown in Fig. 2. The major design areas, such as the evaporator plates, condenser zone, transport lines, and reservoir, are highlighted in the figure and described as follows.

Evaporator Plates

In any CPL system, acquisition of waste heat (or applied heat) at the cold plate provides the motivating force in the system. The primary element

*Manager, Thermal Systems Technology, Astro-Space Division; currently Manager, Thermodynamics, Fairchild Space Company, Germantown, Maryland.

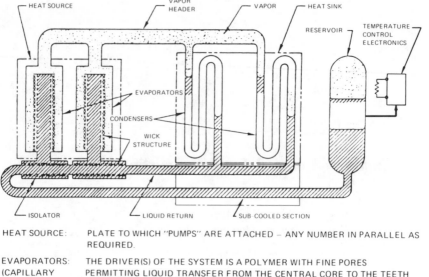

HEAT SOURCE: PLATE TO WHICH "PUMPS" ARE ATTACHED – ANY NUMBER IN PARALLEL AS
                   REQUIRED.

EVAPORATORS: THE DRIVER(S) OF THE SYSTEM IS A POLYMER WITH FINE PORES
(CAPILLARY   PERMITTING LIQUID TRANSFER FROM THE CENTRAL CORE TO THE TEETH
PUMPS)   SURFACE OF A GROOVED TUBE WHERE EVAPORATION OCCURS. THE VAPOR
                   IS THEN TRANSPORTED VIA THE GROOVES TO THE VAPOR HEADER.
                   PRESSURE DIFFERENCES (PUMPING HEAD) IS CREATED BY CAPILLARY
                   ACTION IN THE WICK AND IS SUFFICIENT TO DRIVE LIQUID AND VAPOR
                   AROUND THE CIRCUIT (PISTON ACTION).

VAPOR & LIQUID   COMMON HEADERS – VARIED FOR VARIOUS SYSTEM REQUIREMENTS
HEADERS: (CAPACITY REQUIREMENT, NUMBER OF PUMPS, ETC.)

HEAT SINK/   HEAT EXCHANGER OR RADIATOR. VAPOR IS COMPLETELY CONDENSED
CONDENSERS: WITHIN TUBE (ENTERS VAPOR, EXITS LIQUID).

SUBCOOLED   ENSURES COMPLETE CONDENSATION OF VAPOR BEFORE INPUT TO
SECTION: EVAPORATOR SECTION.

RESERVOIR/   PROVIDES FLUID INVENTORY CONTROL AND THUS REGULATES TEMPERATURES
ACCUMULATOR: (THRUPUT CONTROL).

ISOLATORS: INCORPORATING WICK-ELEMENT TO ISOLATE PUMPS FROM CPL SYSTEM
                   (PREVENTS VAPOR FLOW-BACK CAUSING DE-PRIME).

Fig. 1   Capillary pumped loop system.

in the heat acquisition is the capillary pump. A standard design was
selected for this application and is illustrated in Fig. 3. The pump consists
of a tubular Porex wick that is force-fitted within an axially grooved
aluminum extrusion. Liquid flows along the length of the evaporator
through a passage running through the center of the wick. Liquid also flows
radially through the wick of the fin tips of the extrusion where it evaporates. The vapor then travels along the grooves to the vapor outlet. Direct
contact of the extruded fins with the evaporating liquid on the outer wick
surface creates an inverse meniscus that yields heat-transfer coefficients two
or three times greater than those obtained with conventional, axially
grooved evaporating surfaces.

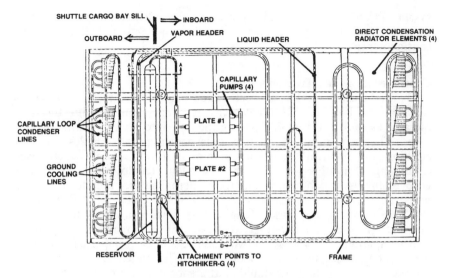

Fig. 2   Capillary flight experiment schematic.

The flight experiment evaporator area consists of two cold-plate assemblies, each containing two evaporator pumps metallurgically bonded into the cold plate. The design uses two cold plates to demonstrate heat load sharing; i.e., the evaporators in either cold plate can operate as condensers. Heat is applied via standard Kapton electrical resistance heaters mounted directly to the cold-plate assemblies via pressure plates.

Liquid insertion to the capillary evaporator pumps is accomplished via stainless steel wicked isolators, consisting of a stainless steel tube (manifold) swagged around a Porex core, that form multiple isolated liquid-return flow paths to each pump. The isolator thermally decouples each flow path and prohibits vapor backflow among the pumps. In this way, failure (deprime) of one or more evaporators does not affect (deprime) the remainder of the

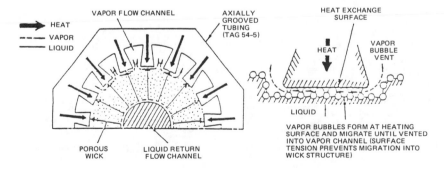

Fig. 3   Capillary pump standard design.

system. The wicked isolator also prevents noncondensable gas bubbles that formed elsewhere in the system from migrating into individual pumps, causing deprime.

Condenser Zone and Transport Lines

The flight experiment condenser design employs a direct condensation radiator (DCR); i.e., the condenser tubes are mounted directly to a radiator as shown in Fig. 4. Dividing the flow into separate paths permits each radiator element to operate independently and enables modular addition and removal. There are a total of four radiator panels, each containing a multipass condenser. The DCR employs flanged, axially grooved tubing, which provides a larger internal surface area for enhanced heat transfer. The tubing is bonded to a single radiator skin. The "paper clip" shape of the condenser design offers a more uniform temperature distribution along the radiator panel and hence increases the radiator's heat-dissipation capability. The two outer legs in each paper clip dissipate the latent heat due to vapor condensation. The center leg provides extra cooling capacity so that the liquid is at a subcooled state before returning to the evaporator section. For further subcooling, a solid fin attaches to the liquid manifold that connects the condenser legs. The exposed radiator surface of the CAPL flight experiment (approximately 1.04 m^2) is covered

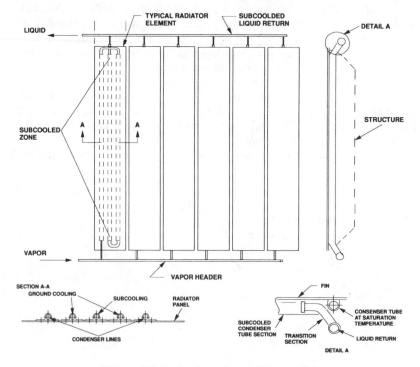

Fig. 4   Direct condensation radiator model.

with aluminized Teflon tape. This includes the top 51 cm of the radiator, which lies above the sill line of the Shuttle bay. This section is designed as a two-sided radiator.

Reservoir

Temperature control for the CAPL flight experiment is accomplished by fluid management using a saturated temperature/pressure-controlled reservoir. The reservoir connects to the CPL liquid side via a reservoir liquid feed line. To ensure proper temperature control, the reservoir must be sized to accommodate the range of liquid displacement within the following two extreme conditions:

1) minimum heat load at the minimum condenser sink temperature, which results in a nearly blocked condenser section; and

2) maximum heat load at the maximum heat sink temperature, which requires all of the condenser to be fully active.

The desired system temperature is achieved by allowing the reservoir to passively radiate to deep space, or a low-temperature environment, with exact temperature control provided via thermostatically controlled electrical heaters. Theoretically, the reservoir's saturation temperature should control the system's operating temperature as long as the following conditions are satisfied:

1) The reservoir is only partially filled with liquid under all conditions.

2) A partially blocked condenser/subcooler section is maintained at all times.

Test results to date have shown that control of the system operating temperature by the reservoir set point is limited only by the sensitivity of the electronic temperature controller.

The flight experiment reservoir volume was conservatively sized to contain enough liquid to completely fill the total system volume at the system's minimum operating temperature prior to startup and accommodate system expansion to the maximum temperature. The reservoir consists of a stainless steel cylinder with internally attached liquid acquisition baffles and a control wick structure that maintains liquid/vapor separation and ensures preferential liquid displacement in 0-g environment. The thermostatically controlled heaters are installed at the bottom of the reservoir near the liquid outlet. Radiator plates attached to the liquid feed line connect the reservoir and the evaporator isolators and provide enough liquid subcooling before the liquid is fed to the isolators.

Flight Experiment Integration and Test

After system integration is completed, the CAPL will be flown and tested aboard the Shuttle to characterize its performance under microgravity conditions. Each test and its objective are described as follows:

System Startup Test

The purposes of this test are to determine conditions under which reliable startup can be achieved and to establish a startup procedure.

Maximum Transport Capability Test

The purposes of this test are to verify the system transport capability of 6000-Watt-m per evaporator and to determine the maximum power that the system can handle.

Rapid Power Cycling Test

The rapid power cycling test is designed to investigate system adaptability to an abrupt change of heat loads from a low power input to a high power input and/or vice versa.

Heat Load Sharing Test

The purpose of this test is to demonstrate the experiment's capability to share heat loads among the evaporators by having some of the evaporators function as condensers.

Pressure Priming Under Heat Load

The objectives of this test are to determine parameters that govern the success of the pressure priming and to establish a reasonable temperature band for the evaporator capture.

Reservoir Set-Point Temperature Variation

The purposes of this test are to verify that the set-point temperature of the reservoir controls the system's operating temperature and that the CAPL can adapt to a sudden change of the reservoir set-point temperature.

Sink Temperature Variation

The objective of this test was to demonstrate the CAPL's insensitivity to large variation in the condenser sink temperature, as long as sufficient cooling capacity is retained.

Diode Function of the Condensers

The CAPL has the capability to automatically act as a thermal diode and therefore shut down the operation if condenser temperature rises above the saturation. This test is designed to verify such a diode function and study the condenser behavior if some of the parallel condensers are temperatures above the saturation whereas others remain at a subcooled state.

Software

Numerous software codes to characterize two-phase fluid heat-transport loops are available or in various stages of development. Of these codes, SINDA 85/FLUINT[4] is the most widely known in the aerospace industry and has the potential to be the industry standard code. The FLUINT capability was added to SINDA '85 by Martin Marietta under NASA-JSC Contract NAS 9-17448.

FLUINT is intended to provide a general analysis framework for internal one-dimensional fluid systems. It can be used alone or in combination with the thermal analysis capability of SINDA '85 and is intended to be to thermal/hydraulic analysis with SINDA '85 is to conduction/radiation analyses. The specific capabilities are described in the following paragraphs.

Arbitrary Fluid Systems
FLUINT is not restricted to any specific geometries or configurations. The program is based on network methods, allowing the user to describe almost any schematic representation.

Fluid Independence
FLUINT may be used to analyze any working fluid with adequately defined properties. Twenty refrigerants are immediately available, and the user may describe the properties of simple gases and liquids with reusable FPROP DATA blocks.

Variable Resolution and Assumption Levels
FLUINT may be used for a wide variety of analyses, ranging from first-order performance approximations to detailed transient response simulations. Spatial resolution may be varied by adding or subtracting network elements, and temporal resolution may be varied by choosing between time-dependent and time-independent types of elements. Furthermore, the performance of fluid components such as pumps can be idealized or as realistic as appropriate to the problem and to the level of known detail.

Steady-State and Transient
FLUINT can be used for either steady-state analyses or for transient analyses. As noted earlier, the fidelity of the transient analysis can be varied from a fully time-independent thermal response (quasi-steady-state) to a fully time-dependent thermal/hydraulic response, depending on the *type* of network element selected.

Single- and Two-Phase Flow
The code handles both single-phase and homogeneous two-phase flow, along with transitions between states. This generality exists not only in the numeric solution, but also in the correlations used for each device or phenomena. In fact, the capability exists to handle capillary devices, where surface tension forces dominate when two phases are present.

Component Models
FLUINT includes general device models for a wide variety of common fluid system components. The user can tailor these models for specific devices. The user can even model new or unique devices by accessing program variables with his own simultaneously executed FORTRAN logic.

The basic building block style network elements were specifically designed for such usage.

Similarity to and Compatibility with SINDA '85

The basic methods in FLUINT contains many analogies to the lumped-parameter style analysis of SINDA '85, which should help the SINDA '85 user learn FLUINT easily. FLUINT may be thought of as a fluids coprocessor to SINDA '85. Analyses may be all-fluid, all-thermal, or both fluid and thermal simultaneously. FLUINT, along with SINDA '85, dynamically sizes the processor for minimum overhead and allows user control and user logic.

FLUINT Limitations

The user should also be aware of current program limitations. These limitations can be divided into two classes. First, there are some limitations that may be circumvented by extensive user logic if the user is sufficiently advanced, but which are not yet available as options. Second, there are certain program capabilities that cannot be appended within user logic, but may eventually be added to FLUINT as growth options. The subsequent paragraphs will briefly mention alternatives for advanced users if any alternatives exist.

Internal Fluid Systems

The program is intended for one-dimensional piping networks, not for external flow problems such as aeroheating.

Working Fluid Choices

Although the code can analyze multiple fluid systems containing different working fluids, no mixtures are permitted—only one fluid substance may exist in any one system. Furthermore, the internal correlations are not applicable to cryogenic and liquid metal fluids. Any single-phase one-component fluid can be analyzed, but the user is currently restricted to 1 of 20 prestored working fluids if a two-phase one-substance analysis is desired because of the extensive property descriptions required.

Flow Conditions

The code is specifically intended for low-speed, viscous flows in ducts. Compressible flow phenomena such as shocks and choked flow are not included, although certain effects such as valve choking can be simulated by advanced users. The program does not make the distinction between static and dynamic temperatures and pressures.

Phase Drift

Within each duct, the liquid and vapor velocities are assumed the same. This effect may be grossly approximated by using two parallel ducts and phase-specific suction on either end, with mass and heat interaction between the phases modeled separately by the user. Interface momentum is not addressed using this approach.

Nonequilibrium States
Within each control volume, liquid and vapor are assumed to be in thermodynamic equilibrium with each other. Superheated vapor cannot coexist with liquid, and subcooled liquid cannot coexist with vapor. Therefore, the addition of a small amount of subcooled liquid to a saturated control volume will cause an immediate pressure and temperature drop. Analogous to phase drift methods, nonequilibrium control volumes may be grossly approximated by subdividing the volume into liquid and vapor spaces, with the heat and mass transfer between the phases simulated separately by the user.

Liquid Compressibility
Liquids are assumed incompressible, meaning that their densities are functions of temperature only. Therefore, the program's capabilities can only be used to approximate water-hammer or column separation effects. However, this code should not be used for pipe-burst or external loading studies, because the correlations employed in the code are not adequate for this purpose.

References
[1]Ku, J., Kroliczek, E., Taylor, W., and McIntosh, R., "Functional and Performance Test of Two Capillary Pumped Loop Engineering Models," AIAA Paper 86-1248, June 1986.

[2]Ku, J., Kroliczek, E., Butler, D., Schweikhart, R., and McIntosh, R., "Capillary Pumped Loop GAS and Hitchhiker Flight Experiments," AIAA Paper 86-1249, June 1986.

[3]Chalmers, D. R., Fredley, J. E., Ku, J., and Kroliczek, E. J., "Design of a Two-Phase Capillary Pumped Flight Experiment," Society of Automotive Engineers Paper 881086, July 1988.

[4]Users Manual—SINDA 85/FLUINT Version 2.1, Martin Marietta, MCR 86-594, Nov. 1987.

Chapter 3. Startup Thaw Concept
for the SP-100 Space Reactor Power System

Startup Thaw Concept for the SP-100 Space Reactor Power System

A. Kirpich,* A. Das,† H. Choe,‡ E. McNamara,§ D. Switick¶
General Electric Company, Princeton, New Jersey
and
P. Bhandari‖
Jet Propulsion Laboratory, Pasadena, California

Introduction

The SP-100 Ground Engineering System (GES) Program is addressing the development and validation of technologies needed to implement space reactor power systems. Under contract with the Department of Energy, General Electric is playing a key role in the GES Program in conceptualizing and defining space reactor power system (SRPS) designs capable of producing electrical power in the range of 10–1000 kWe. Studies over the past 5 yr have converged on SRPS concepts with the following attributes:

1) fast spectrum, liquid-metal-cooled reactor operating at 1300–1400 K;
2) pumped lithium heat transport;
3) conductively coupled thermoelectric power conversion; and
4) pumped lithium heat removal to waste heat radiator consisting of potassium heat pipes brazed to supply and return fluid ducts.

The design of a 100-kWe reference flight system that incorporates the preceding attributes was recently completed and will serve as the basis for

*System Design Manager, SP-100 Program, Astro-Space Division.
†System Analyst Manager, SP-100 Program, Astro-Space Division.
‡Lead Engineer, Primary Heat Transport Subsystem Design, SP-100 Program, Astro-Space Division.
§Lead Engineer, Heat Rejection Subsystem Design, SP-100 Program, Astro-Space Division.
¶Nuclear System Engineer, SP-100 Program, Astro-Space Division.
‖Member Technical Staff.

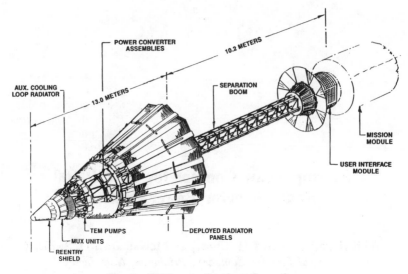

Fig. 1   RFS operational configuration.

GES Program development and validation planning. Figure 1 provides an overview of the concept in its deployed configuration and shows where a hypothetical mission module payload would be located relative to the reactor and power conversion equipment. A shield behind the reactor attenuates the nuclear radiation within a 17-deg half-cone region. This region delineates the boundaries within which all SRPS equipment and components are contained. Further attenuation at the payload location is achieved by the separation boom, which reduces radiation levels according to a reciprocal R^2 relationship. With the boom collapsed and the radiator panels (12) folded forward in the stowed position, the SRPS fits within a 7-m length of the National Space Transportation System (Shuttle) bay.

A simplified hydraulic schematic of 1 of 12 modular segments is shown in Fig. 2. The modular approach results in standardized building blocks that permit scaling adjustments over a broad power range in increments of

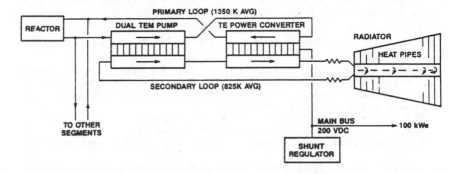

Fig. 2   Simplified functional schematic.

about 8 kWe. Lithium is used as the fluid because of its density, specific heat, and vapor pressure properties. Its frozen condition during launch provides safety advantages in that any launch-induced piping rupture will not endanger crew or equipment.

After insertion into the required orbit, the lithium distributed through the system is thawed by a heat pipe network that derives its heat from the reactor. Upon full thaw the primary loop high-temperature lithium is circulated through the reactor (which serves all 12 segments) and is routed to the inlet of a thermoelectric-electromagnetic (TEM) pump. This pump delivers the fluid to the hot side of a thermoelectric power converter and then returns the fluid to the reactor. Heat is given up as the fluid passes through thermoelectric elements in the power converter and TEM pump. In the process, some of the heat is transformed to electricity and the remaining heat is transferred to a secondary lower-temperature lithium loop for transport to a heat pipe radiator and rejection to space. Electricity developed within the power converter is combined with that from the other segments and delivered as the system output (100-kWe rating) at 200 V. A shunt regulator maintains voltage within regulation limits by dissipating excess power not demanded by the user load. Electricity developed within the TEM pump operates in conjunction with a self-induced magnetic field to produce the hydraulic pumping force. The TEM pump is dual-acting by virtue of producing hydraulic pumping for both the primary and secondary loops within a single pump body. The primary loops of 12 pumps (1 per segment) operate from a single fluid space, whereas the secondary loops are hydraulically independent.

The background provided earlier serves as an introduction to a broader discussion of the thaw concept that has evolved during the definition of the SP-100 reference flight system (RFS). The following thaw concept description has been tailored to the specific needs of the 100-kWe RFS; however, the methods employed are applicable to a broad range of power ratings for space reactor power systems of the type described.

Coolant Selection

The need for thaw derives from the basic selection of lithium, which has a melting temperature of 454 K. Lithium has been selected over other candidates because of its low vapor pressure at the intended reactor temperature (1300–1400 K), its high specific heat, and its relatively low density. Table 1 compares the characteristics of lithium with other potential coolants. Particularly significant are its relatively low pumping power, largely resulting from its high specific heat, and its low vapor pressure, which minimizes the structural requirements for vessel walls and piping.

However, the selection of lithium carries the burden of thaw, a process that involves considerable interplay among the various thermal subsystems that make up the SRPS.

The thaw phenomenon involves various changes in thermophysical properties. The compilation of Table 2 provides some of the important properties of lithium needed in developing the thaw system design. Figure 3

Table 1   Physical properties of potential coolants for SP-100

	Li	Na	NaK	K
Density, kg/m³ (300 K)	530	970	890	860
Melting point, K	454	371	260	337
Boiling point, K	1606	1155	1058	1033
Vapor pressure, kPa				
1220 K	2.1	172	276	483
1350 K	8.3	427	772	931
Electrical resistivity, $\mu\Omega$-cm (1350 K)	49	53	140	125
Thermal conductivity, w/mK (1350 K)	61	42	22	23
Viscosity, centipoise	0.161	0.13	0.11	0.084
Specific heat, kJ/kg K	4.17	1.29	1.06	0.833
Relative convective film coefficient[a]	1.0	0.85	0.60	0.65
Relative pumping power[a]	1.0	8.5	17	42

[a]Normalized to lithium.

provides volumetric expansion data that are important in establishing lithium volume requirements and accumulator design and in assessing structural loading conditions.

Thaw Concept
The basic strategy for system thaw is to extract heat from the reactor and deliver it to the various thermal subsystems by several means. Initial thaw heat is provided by heat pipes that are attached to the reactor vessel. They deliver sufficient heat to thaw the piping, circulating pumps, and power converters.

Table 2   Thermophysical properties of lithium

Temperature T, K	Density ρ, kg/m³	Specific heat C_p, kJ/kg K	Thermal conduct k, w/mK	Dynamic viscosity μ, Ns/m²	Heat of vaporization h_{fg}, kJ/kg	Surface tension σ, dyne cm
453.7	523	4.30	42.8	6.00×10^{-4}	22556	397.8
500	514	4.34	43.7	5.31×10^{-4}	22485	391.2
600	503	4.34	46.1	4.26×10^{-4}	22340	377.2
700	493	4.19	48.4	3.58×10^{-4}	22172	363.2
800	483	4.17	50.7	3.10×10^{-4}	21989	349.2
900	473	4.16	52.9	2.75×10^{-4}	21751	335.2
1000	463	4.16	55.2	2.47×10^{-4}	21471	321.2
1100	452	4.15	57.6	2.25×10^{-4}	21156	307.2
1200	442	4.14	59.8	2.07×10^{-4}	20818	293.2
1300	432	4.16	62.1	1.92×10^{-4}	20466	279.2
1400	422	4.19	64.0	1.80×10^{-4}	20110	265.2
1500	411	4.20	66.5	1.69×10^{-4}	19757	251.2

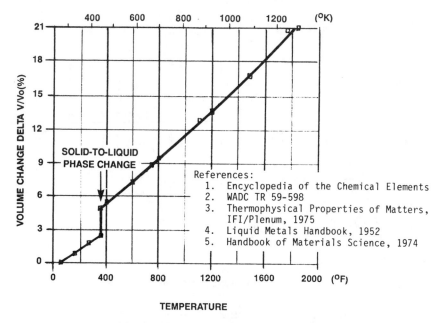

Fig. 3   Lithium expansion characteristics.

Once circulation is established, the flowing lithium brings heat from the reactor that is used to thaw the liquid ducts and heat pipes of the large heat-rejection radiator.

Prethaw Considerations

The thaw concept for the SRPS discussed herein is one in which the hydraulic components and piping are filled with lithium, which is frozen in place during prelaunch preparations. Voids are placed at strategic locations in the hydraulic circuits to accommodate thaw expansion. The distribution of frozen lithium and voids throughout the system represents the state of the SRPS from prelaunch until thaw is initiated after orbit insertion. This distribution of frozen lithium and voids applies to the reactor, primary loop piping, volume expansion accumulators, TEM pumps, power converter hot- and cold-side heat exchangers, secondary loop piping, and radiator fluid ducts. The prelaunch fill procedures involve purging the containment vessels and piping and filtering the circulated lithium to ensure ultraclean conditions and virtual elimination of contaminants. Voids are created in specific locations by elevating the zone manometrically to reduce pressure levels to those below the corresponding vapor pressure of the lithium. This causes the fluid to break away, resulting in void formation. This procedure is accomplished in conjunction with generating a nucleation site, which refers to the need to create a void "seed" because of the high wettability of lithium and its resistance to being separated from the containment wall surface. Nucleation sites may be generated thermally, by appropriate

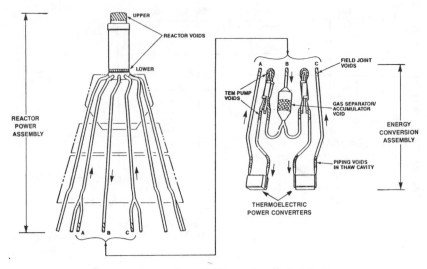

Fig. 4   Primary loop lithium and void distribution.

surface irregularities, or by the introduction of a nonwettable material at the intended site. Freezing of localized plugs on one side of the intended void may be employed as a means of ensuring void location.

The distribution of frozen lithium and void distribution is illustrated in Fig. 4, which shows the elements and piping of the GFS primary loop. Voids are located in both upper and lower regions of the reactor, and their size is determined by the lithium expansion that will occur before flow is initiated. (A later discussion will cover the startup sequence.) Points, A, B, and C define the interface between the reactor power assembly and one of six identical energy conversion assemblies. Voids are introduced at these points as a consequence of the separate fill procedures for these assemblies and the sequence of system buildup.

Within each energy conversion assembly, voids are introduced near the TEM pumps and at several locations along the primary pipes. The need for these voids has been determined as a result of thermal analyses, which will be discussed later. A large void is also introduced within the gas separator/ accumulator unit, which is backfilled with helium and, upon thaw, will serve as a gas spring accumulator for the primary loop.

Voids within the secondary loops are of a similar nature; their need has also been determined by thermal analyses. Specific voids are introduced within the flexible joints of the ducts to the deployable radiator panels. These voids permit deployment of the panels before thaw is initiated. This is necessary to ensure that certain equipment has an adequate view of space to avoid overheating. The resulting sustained high level of flow-convected heat from the reactor is used to thaw the liquid ducts to which the radiator heat pipes are attached.

Thaw initiation is preceded by boom extension, radiator panel deployment, and the release of several mechanical locks that provide structural

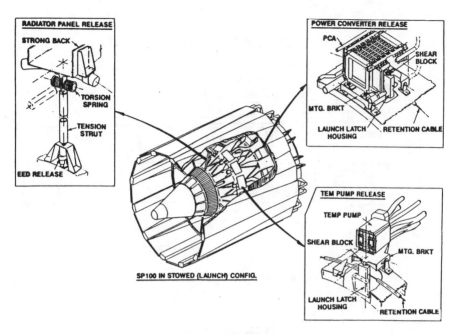

Fig. 5   Launch lock releases.

support during the launch phase. These prethaw steps are necessary because the temperatures associated with thaw would be injurious to electronic equipment, which in the stowed configuration is contained within the radiator panel enclosure. Boom extension avoids this problem for the electronics contained within the user interface module, and panel deployment results in improved space view factors for reactor control multiplexer units mounted aft of the radiation shield. As noted earlier, prethaw deployment of the radiator panels requires that the secondary loop flex joints at the panel hinges be voided of frozen lithium.

Prethaw release of mechanical locks for the TEM pumps, gas separator/accumulators, and power conversion assemblies is needed to accommodate the differential expansion of primary and secondary loop piping that will occur as the system heats up. Provisions for such releases are shown in Fig. 5. The general approach is to maintain component locations by means of circumferential tension cables that are released by cable cutters after the powered flight phase.

Thaw Provisions

The thaw system schematic is illustrated in Fig. 6 for 1 of 12 similar segments. For clarity, the schematic has been simplified, particularly with regard to the reactor hydraulic and thaw heat pipe (THP) connections. The top sketch portrays the segment primary loop including the TEM pump and hot-side heat exchanger (HSHX), which delivers heat to the power

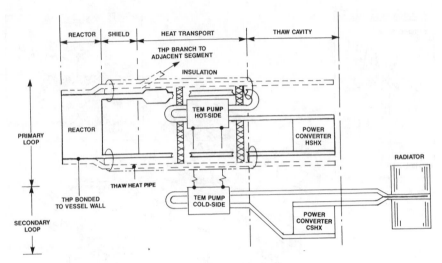

Fig. 6   Thaw system block diagram.

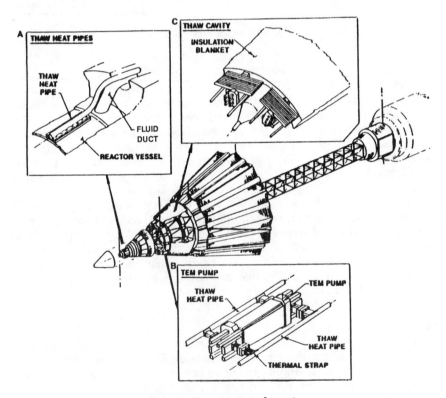

Fig. 7   Thaw system elements.

converter. The sketch also shows the THP's bonded to the reactor vessel and conducting straps to the TEM pump.

The bottom sketch of Fig. 6 shows the secondary loop, including the cold-side duct of the TEM pump, the cold-side heat exchanger (CSHX), which removes heat from the power converter, and the radiator.

Some of the key features of the thaw concept are shown in Fig. 7 and will be referred to in the following discussion. Thaw is initiated within the reactor core by a programmed addition of reactivity through adjustment of the control reflectors. Reactivity additions are initially small so that core temperature increases are gradual to avoid severe temperature gradients and any tendency toward overstress.

At an intermediate core temperature level, the THP's attached to the external vessel wall (Fig. 7a) become operative. Each of the 12 THP's delivers heat by radiation coupling to the six supply and six return transport lines to and from the reactor vessel. Multilayer insulation (MLI) encloses each THP/transport line pair. The THP/transport line pairs trace from the reactor vessel along spiral grooves located on the periphery of the conical radiation shield. At a point just aft of the shield, the vessel-wall THP is mechanically spliced to two branching THP's, which continue further aft and deliver heat to the TEM pumps and to the thaw cavity enclosing the power converter assemblies and primary and secondary loop piping.

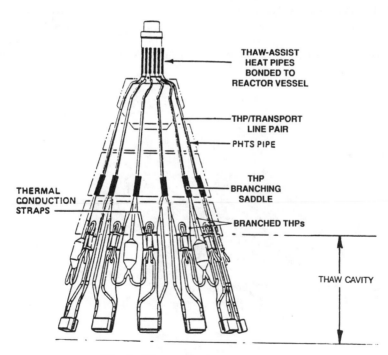

Fig. 8 Thaw-assist heat pipe layout.

Figure 8 provides a broader view of the THP network showing the bonded connections to the reactor vessel, the tracing of the THP's along the primary loop line from the reactor, and the branching saddles to which THP pairs are spliced and routed aft to the thaw cavity. Figures 7b and 7c show the principal features of the thaw design associated with the TEM pumps near the entry to the thaw cavity. Two THP's, each representing a branch from their respective reactor vessel THP's, are routed laterally along each TEM pump. Conductive straps (Fig. 7b) deliver heat from the THP's to the four primary loop entry points to the TEM pump. This heat not only thaws the pump by longitudinal conduction from the pump ends but also delivers sufficient heat to the TEM pump thermoelectrics (TE's) to initiate pumping (i.e., it causes a temperature gradient across the TE's to produce pumping power). Figure 7c shows a small heat pipe radiator brazed to the external cold-side duct of the pump. Its purpose is to establish a sufficient temperature difference across the TE's so that flow will be initiated. Once hot reactor fluid is transported to the pump, its pumping capability will increase and the THP-conductive strap heat-delivery system will have performed its pump startup function.

Flow cannot proceed until the full primary loop has been thawed. Figure 6 shows that this requires thaw of the power converter and piping enclosed within the thaw cavity. As shown in Fig. 9, the thaw cavity is a thermally insulated annular enclosure that surrounds the power converter and piping. Thaw of the cavity enclosed components is achieved by the THP's extended into the cavity which deliver heat radiatively. As shown on Fig. 6, thaw of the power converters is accomplished by the THP's that radiate to the uninsulated cold-side heat-exchanger manifolds. The intercepted heat is conducted inward from the manifolds to thaw the lithium contained in the

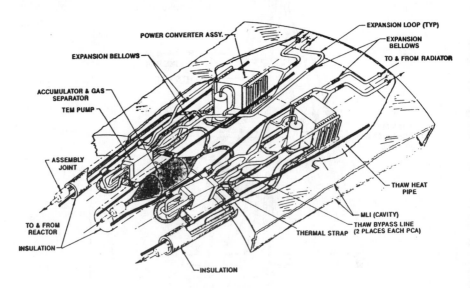

Fig. 9 Thaw cavity.

cold-side heat exchangers. Thaw within the cavity is reversed in the sense that the cold-side heat exchangers are thawed first, and they, in turn, thaw the hot-side heat exchangers.

With thaw of the cavity-enclosed primary loop elements, the pump can circulate hot reactor fluid through the system, permitting thaw to proceed further. Several concerns at this stage pertain to the abrupt increase in flow that occurs with the final removal of frozen blockage in the primary loop. The first pertains to the change in reactivity resulting from circulating the initially cool fluid of the power converter back to the reactor; the second pertains to the possible thermal shock of initially circulating the hot reactor fluid into the relatively cool power converters. Both of these concerns were addressed during early transient analyses which indicated that flow restriction immediately following primary loop thaw completion would be desirable to mitigate the reactivity and thermal shock effects. Once a closer match of reactor and power converter temperature levels is achieved, the flow restriction could be removed. Recent analyses have confirmed the nature of these concerns and the necessity for initial flow restriction. Figure 10 shows a simple scheme that has been devised and analyzed for the required function. A small-bore tube is installed in parallel with the primary line located between the TEM pump and power converter inlet manifold. This tube is located within the thaw cavity and is designed to thaw before the primary line itself. This is accomplished by using less insulation than is used on the primary line itself. The thaw of the primary line must also occur after the power converter is thawed. With these design conditions satisfied, the small bypass line allows a relatively small flow to be established through the power converter, permitting thermal equilibrium to be established gradually with the reactor without severe reactivity and thermal shock effects. With complete thaw in the main line, resulting transient conditions will be considerably relieved.

Up to the time that the components within the thaw cavity are thawed, the reactor bulk temperature is maintained at a reduced temperature limit. This limit is principally imposed by the material properties of titanium used in the secondary loop components (pumps, power converter, piping). Once flow is initiated, the heat exchanger and piping of the secondary loop will approach the reduced temperature level since the means of heat rejection are limited at this stage. That is, the radiators have not as yet been thawed; therefore, heat flows from the primary to secondary loops (through the power converters and TEM pumps) are small with correspondingly small temperature drops.

Secondary loop ducts emerge from the aft face of the thaw cavity and connect to the flexible joints of the radiator panel hinge line. The flexible joints in turn connect to supply and return ducts routed longitudinally over the entire radiator panel length. Transverse heat pipes are bonded directly to the duct walls. Thaw of the emerging secondary loop ducts, the flexible joints, and the longitudinal ducts is accomplished by a unique arrangement in which supply and return fluid is interchanged, resulting in a fully progressive method of thaw. The thaw front proceeds from the aft face of the thaw cavity, through the flexible lines, and over the length of the ducts.

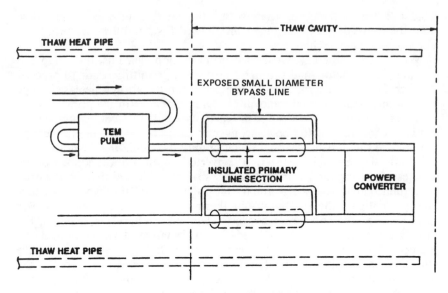

Fig. 10 Startup flow restrictor.

Thaw of the attached transverse heat pipes occurs as the secondary loop thaw front progresses along the supply and return ducts.

It was noted earlier that the flexible joints are voided to permit easy deployment before thaw is initiated. When the thaw front discussed earlier just reaches the flexible joint voided region, the molten fluid will fill the void as a result of the pressure and volume buildup in the secondary loop accumulator located near the TEM pump. The volume buildup occurs as a result of the fluid thermal expansion.

Control Strategy

The reactor control strategy is based on achieving autonomous thaw of the system consistent with the following physical events and objectives:

1) Achieve initial reactor thaw in a nonflow environment:

    a) Establish reactor neutronic criticality and thermal power production within safe power and temperature overshoots using open-loop reflector control.

    b) Establish closed-loop feedback control based on reactor vessel wall temperature, within predetermined thaw rates limits.

    c) Maintain reactor vessel wall temperature at reduced levels until thaw of other components is completed via the thaw heat pipe and conduction strap network.

2) Maintain reactor temperatures within safe limits during primary loop flow initiation transient.

    a) Confirm flow establishment in all primary loop branches.

    b) Maintain reactor vessel wall temperature at a reduced level as the heat source to thaw secondary loops.

3) Bring reactor power up consistent with using reactor outlet temperature for control.
4) Bring reactor temperature up to full-power condition.

Thermal Analyses

Thermal analysis methods are basic to thaw system design for predicting thaw behavior. The analytical results provide both spatial and temporal thermal distributions, indicate the nature of thaw progression, and provide the basis for estimating the overall thaw time. Presented in this section are some of the thermal analysis results developed for the 100-kWe SRPS described herein. These results are representative of the type of thermal analysis that would be undertaken in developing a thaw concept and are presented in this section in the chronological order in which the actual transient takes place.

Reactor Thaw

Thaw of the reactor is the critical first step in achieving overall thaw and is accomplished by bringing the reactor to a critical state through reactivity additions resulting from safety rod withdrawal and the movement of external reflector segments. Controlled movement of these elements results in the generation of prescribed levels of thermal generation by the nuclear fission process. From a thaw standpoint, the initial heat addition must be carefully metered because of the spatially nonuniform distribution of heat generation. In general, heating will be highest at the center of the reactor core. Coupling this fact with the expansion characteristics of melting lithium (see Fig. 3), a condition arises in which an entrapped zone of liquid lithium presses outward against an enclosing shell of solid lithium. This may result in plastic deformation of the solid lithium itself with possible deleterious effects on surrounding reactor structural elements. Study of this issue indicates that serious effects can be mitigated by incorporating strategically located void zones within the reactor and by reduced rates of heat addition. Solid lithium near its melting point is somewhat gummy in character and can be extruded through and around intermittent solid barriers as would be represented by the reactor internal fuel pins and mounting hardware. Studies to date indicate the feasibility of taking advantage of the extrudability characterisitc in avoiding any possibility of structural damage.

The nominal heat addition profile of Fig. 11 is designed to accomplish reactor thaw in a benign fashion according to the constraints noted earlier. The initial period of heat addition is at a sufficiently low rate to avoid excessive stress levels. A dwell period is then introduced that results in the leveling of internal temperatures. This is followed by increased rates that elevate the reactor to intermediate temperatures, adequate to light-off the externally mounted thaw heat pipes. Figure 12 shows the internal reactor temperature-time heating history resulting from this startup strategy.

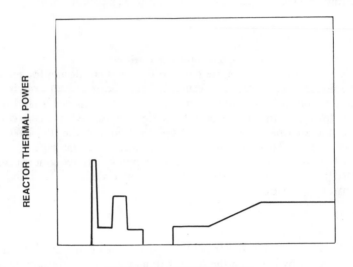

Fig. 11   Nominal reactor power-time history.

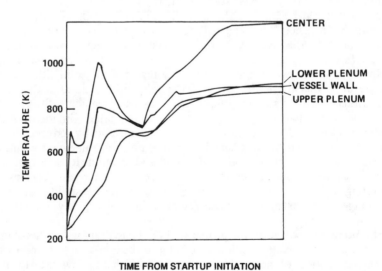

Fig. 12   Reactor temperature-time history.

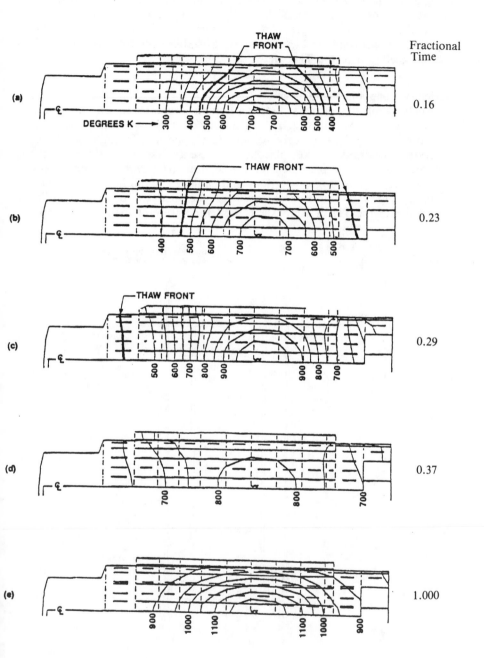

Fig. 13   Reactor thaw progression.

Reactor thaw profiles, shown in Fig. 13, illustrate the thaw front within the reactor shortly after initial thermal power generation. Figures 13b–13d show the thaw front progression in the core. Finally, Fig. 13e shows the internal temperature profiles for the quasistable condition during which heat is delivered to the THP network.

Primary Loop and Pump Thaw

Analyses of the primary loop and pump were based on the region schematically shown in Fig. 14. An intermediate reactor vessel wall temperature was used as the driving boundary condition for lighting-off the potassium heat pipes. A key objective of these analyses were to explore the appropriate size and placement of conduction straps and the pump thaw radiator to ensure progressive thaw and sufficient pump head for startup.

Figure 15 shows a typical result of the analyses performed. With operation of the THP's established, the temperature of the reactor supply

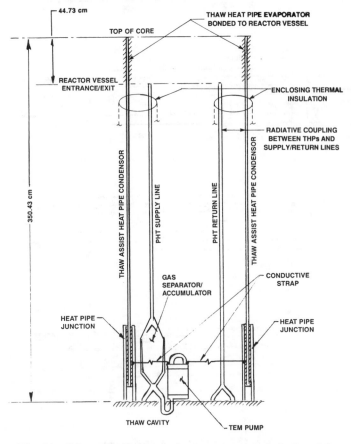

Fig. 14   Schematic of primary heat-transport analytical model.

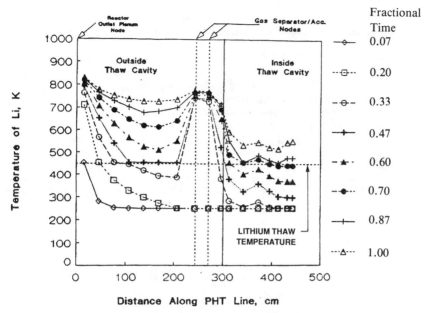

Fig. 15   Temperature history in PHTS supply line.

and return lines increases as indicated by the temperature-time profiles shown. As stated earlier, an important issue is to ensure that thaw occurs progressively; i.e., thaw moves forward in a defined manner with the molten fluid behind the thaw front in direct communication with an expansion space. The profiles of Fig. 15 may appear to contradict this objective in the sense that higher temperatures occur downstream as the thaw front (which occurs at 450 K) advances along the lithium lines. The higher temperatures are the result of the conductive links provided by metal straps that connect the THP's to the TEM pump hot-side ducting. An expansion space for this higher-temperature region is provided by TEM pump voids and by gas separator/accumulator units that are in close proximity (see Fig. 4). Hence, an additional thaw front can start in a direction opposite to the original thaw; thus, the arrangement satisfies the requirements for having the thaw front in fluid communication with an expansion space.

Another key issue concerns the manner in which pumping of the TEM pumps is initiated. The startup dynamics of the pump itself requires that an adequate temperature difference be established across its thermoelectric elements as a precondition for pump startup. The thermal resistances at every point along the heat path from reactor to the pump heat input must be carefully considered in arriving at a satisfactory design. Careful modeling and thermal analysis are indispensable to this critical process.

Cavity Thaw

Cavity thaw is similar to primary loop thaw in that thaw heat is delivered by the THP's that are extended into the thaw cavity region.

Table 3 Cavity thaw timeline

Time	Event
0.0[a]	Second (cavity) thaw heat pipes begin to thaw
0.8	Heat pipe in cavity light off (at oper. temp.)
1.2	Second thaw heat pipe completely lit
1.8	TEM pump primary lithium lines thawed
2.2	TEM pump secondary lithium lines thawed
2.6	HRSS piping in cavity thawed; cold side of PCA thawed
2.8	All lithium in PCA thawed
3.2	PHTS lithium in cavity thawed; all lithium in cavity thawed

[a]Initiation of these events occurs approximately 3.5 h after the receipt of the startup command.

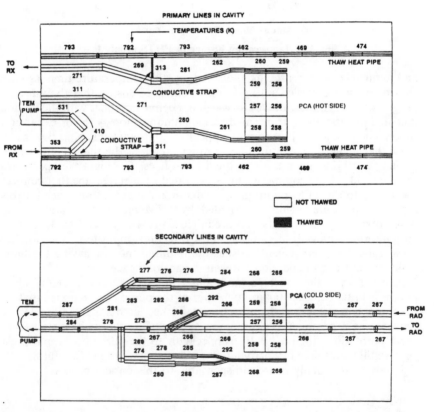

Fig. 16 Cavity temperature/thaw map: time from start of cavity heat pipe thaw = 1 h.

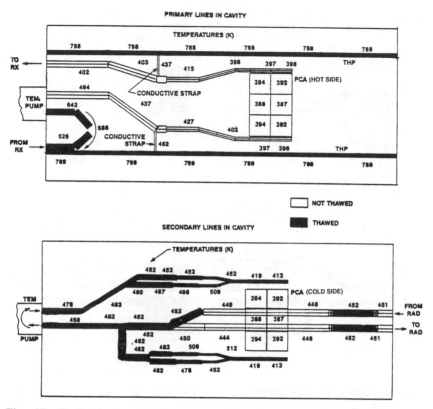

Fig. 17 Cavity temperature/thaw map: time from start of cavity heat pipe thaw = 2 h.

Table 3 shows a timeline of the cavity thaw progression. The thaw is progressive through the primary heat transport system (PHTS) lines, which are wrapped in insulation with the THP's. At 0.8 h, the melt zone of the THP has reached the cavity and begins to provide heat energy to the cavity. The TEM pump thaws at 2.2 h, the heat-rejection piping in the cavity is thawed by 2.6 h, the power converter is thawed by 2.8 h, and the primary heat-transfer lines in the cavity thaw at 3.2 h. Figures 16–18 show nodal thermal maps for the times indicated. A careful study of these maps reveals considerable information about the thaw process and the need for certain thaw features. Some of these points are noted as follows. On the secondary side, the power converter assembly (PCA) headers are the last to thaw. For the primary side the last area of the primary lines to thaw were the sections of pipe midway between the TEM pump and PCA in the PCA supply line and midway between the PCA primary return line and the section of the primary return outside the cavity. This area is well insulated; it thaws from radiation through the MLI and from heat conduction from either

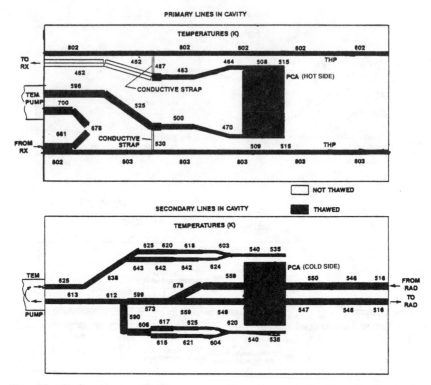

Fig. 18 Cavity temperature/thaw map: time from start of cavity heat pipe thaw = 3 h.

side. To improve thaw in this area, conductive straps could be added from the THP to the midway section of primary line mentioned previously. Examination of the thermal maps indicate that thaw is nonprogressive in some of the piping. Insulating the pipes in such zones will impede radiative heat flow and should result in fully progressive thaw from voided locations. As noted earlier, such mapping is invaluable for understanding thaw behavior and for incorporating any necessary special features.

Radiator Thaw

Radiator thaw commences upon complete thaw of the components contained within the thaw cavity. This is necessary to unblock the passages of the secondary loop leading up to the supply and return ducts of the radiator panels. Upon the development of a pumping head in the secondary loop, thaw progression takes place along the radiator ducts by virtue of the unique fluid interchange arrangement employed. Analytical results predict a radiator thaw time of 8.7 h, including a 1.5-h period for thaw of the flexible joint bellows.

Startup Performance

An end-to-end startup analysis of system thaw was performed using the thermal analyses described previously and combined into an overall simplified simulation. The results, described later, present the time dependence of key performance parameters including reactor power and temperature, fluid temperatures and flow, electrical power output and voltage, and radiator temperatures.

Figure 19 shows reactor thermal power and core temperature during startup. The thermal power spike at 5 h is the result of the primary loop thaw breakthrough and resulting reactivity response to relatively cold fluid return from the power converters. The flow restrictor bypass tube mitigates the severity of this transient. Reactor thermal power gradually increases as more and more heat-rejection capability is established.

Figure 20 shows the key system temperatures and flows that develop during the course of thaw. Primary loop temperature rises to reactor temperature (950 K) upon thaw breakthrough at $t = 5$ h. Secondary loop temperatures gradually decrease as more and more of the radiator is thawed. The difference of primary and secondary loop temperature is directly proportional to reactor heat flow (see Fig. 21) and is determined by the thermal conductivity characteristics of the power conversion system. Figure 21 shows the power generation and voltage buildup resulting from the heat flow through the power converters.

The final plot, Fig. 22, shows the thaw history of the root, midpoint, and tip of the radiator. At 14 h, the return passage at the extreme end of the

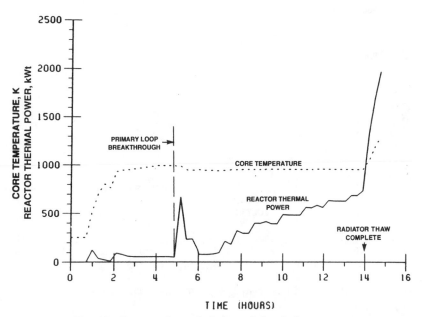

Fig. 19 Reactor thermal characteristics during startup.

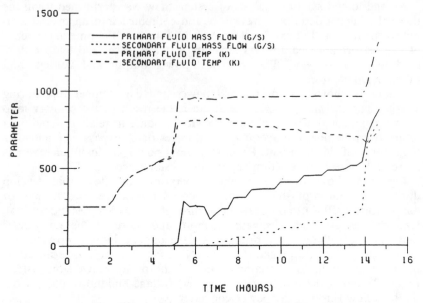

Fig. 20   Reactor transport characteristics during startup.

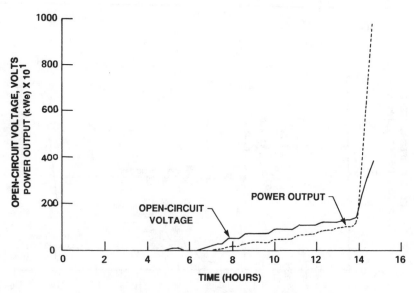

Fig. 21   Power converter performance during startup.

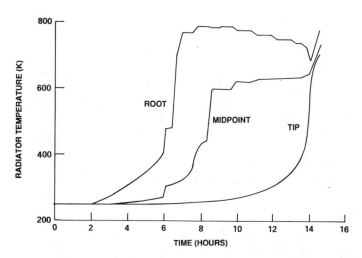

Fig. 22 Radiator temperatures during startup.

radiator has been thawed, permitting a significant increase in secondary loop flow. The sharp rise in performance is attributable to this break-through. At that point the reactor temperature is raised to the 1300 K level and output performance increases accordingly, as shown on all of the figures for $t = 14$ h.

Thaw Volume Progression

An important aspect of system thaw is the time determination of lithium and void volume states at various locations in the system. This is necessary to ensure that entrapped zones of molten lithium are avoided, since their inability to expand freely may lead to structural damage. With respect to the design described herein, the overall volume expansion problem was addressed after the completion of the thermal performance analyses discussed previously. Those analyses identified the potential regions of entrapment simply by those having temperature levels in excess of the lithium melting point (454 K) and for which no fluid communication existed with a free expansion space. For such zones initial voids were introduced as described earlier in the section concerning prethaw consideration.

The series of diagrams presented in Fig. 23 provide a graphic representation of volume states from the initial prethaw state and provide an insight into the problems likely to be encountered and possibilities for their resolution. Besides the emplacement of voids, the use of accumulators provides an important means for resolving fluid expansion difficulties. Both methods of volume expansion accommodation are depicted in the diagrams. Further discussion is provided later for each of the states illustrated in Fig. 23. Volume conditions in both the primary and secondary loops are considered.

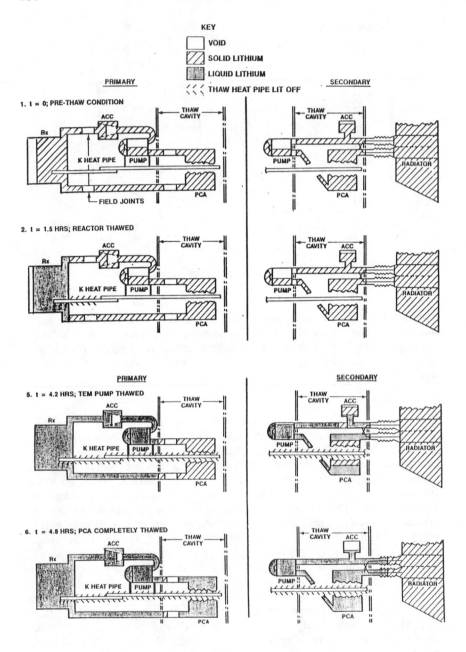

Fig. 23   Thaw sequence.

(Figure continued on next page.)

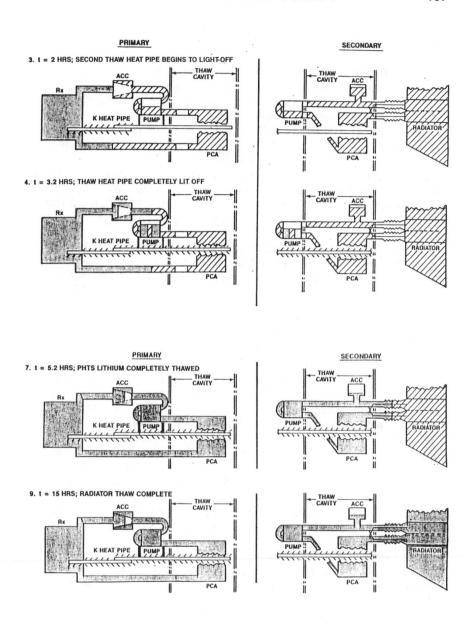

PRIMARY

3. t = 2 HRS; SECOND THAW HEAT PIPE BEGINS TO LIGHT-OFF

SECONDARY

4. t = 3.2 HRS; THAW HEAT PIPE COMPLETELY LIT OFF

PRIMARY

7. t = 5.2 HRS; PHTS LITHIUM COMPLETELY THAWED

SECONDARY

9. t = 15 HRS; RADIATOR THAW COMPLETE

Fig. 23 (cont.)   Thaw sequence.

1) Prethaw condition: The diagram depicts the voids as initially emplaced. In the primary loop, voids are located within the reactor, in the field joints, in the gas separator/accumulator (GSA), in the TEM pump, and in the TEM pump and flexible joints to the radiator. Not shown are voids that may be needed for the secondary piping within the thaw cavity.

2) Reactor thaw: With the reactor raised to an intermediate temperature level, the internal void is reduced in size but not totally eliminated since the supply and return lines are still to be thawed. The potassium thaw heat pipes are lit-off and thaw of the primary loop lines is proceeding. No change has occurred within the secondary loop.

3) Branched thaw heat pipes light-off: The supply and return lines are thawed and fluid communication is established with the GSA. Expansion of the helium within the GSA backfills the remaining void within the reactor. The branched heat pipes light-off, permitting thaw of the TEM pumps to proceed.

4) Branched heat pipes completely lit-off: With the THP's fully operative and delivering heat, TEM pump thaw is initiated by the conductive straps joining the THP's to the pump. Conduction from the hot side of the TEM pump through its thermoelectric elements initiates thaw of the TEM pump secondary loop passages.

5) TEM pump fully thawed: Voids in the TEM pump primary ducts disappear as fluid communication with the GSA is established. The THP's in the thaw cavity thaw the secondary lines (since the primary lines are insulated, they thaw later). Communication with the secondary loop accumulator is not fully established; therefore, by design a small void still remains in the TEM pump. During this phase, thaw of the PCA CSHX's is initiated by radiation from the THP's to the CSHX manifolds. The HSHX's manifolds are insulated; therefore, the CSHX's are thawed first.

6) PCA completely thawed: With the CSHX's completely thawed, the HSHX's are thawed by conduction through the thermopile. Because of primary piping insulation, sections are still frozen and voids are needed to permit free expansion of the PCA HSHX lithium. On the secondary side, the accumulator is thawed and the lithium up to the flexible joints is thawed. Under this condition the accumulator discharges fluid into the flexible joint voids. Some of the discharged fluid is refrozen upon contact with the cold flexible joint.

7) PHTS completely thawed: The primary lines and PCA's are thawed, permitting the possibility of pumped circulation. With a temperature difference established by the pump starter radiator, pumped circulation is initiated.

8) Radiator completely thawed: With continuing pumped circulation, thaw progression occurs along the radiator duct with the radiator heat pipes thawed by conduction. With expansion of the radiator duct fluid, the secondary accumulator is backfilled to an intermediate level.

Conclusion

In this monograph we have described the thaw concept for a space reactor power system that employs lithium as the circulant for both primary heat-transport and secondary heat-rejection fluid loops. The complex nature of the thaw process demands a detailed level of treatment. We hope that the description and analytical methodologies presented will convey the intricate nature of the thaw problem.

We also wish to point out that the solution as presented represents one of many possibilities considered during the early stages of the system design. This particular solution was chosen for further detailed analysis because it was in our judgment the most reasonable and workable. Further studies of the SP-100 reference flight system thaw design are planned and will consider other viable alternatives.

As a final note, we believe that thaw can be successfully achieved with the application of relatively conventional technology. The greatest challenge of thaw system design relates to understanding the various physical states throughout the system, both spatially and temporally, and providing adequate margins that will avoid structural or thermally induced damage.

Chapter 4. Calculational Methods and Experimental Data for Microgravity Conditions

Calculational Methods and Experimental Data for Microgravity Conditions

William J. Krotiuk*

General Electric Company, Princeton, New Jersey

and

Z. I. Antoniak†

Battelle, Pacific Northwest Laboratories, Richland, Washington

Nomenclature

A = flow area
A_H = heat-transfer surface area
C_p = specific heat at constant pressure
d = hydraulic diameter
D = diffusivity
f = friction factor
g = gravitational acceleration
$\boldsymbol{g}$ = gravitational acceleration vector
H = heat-transfer coefficient
h = enthalpy
k = thermal conductivity
L = length
P = pressure
P_w = wetted perimeter
Q_w''' = wall heat flux per unit fluid volume
q = fluid-fluid conduction heat flux
q_I''' = interfacial heat flux per unit volume
T = temperature
t = time

*Principal Member Technical Staff, Astro-Space Division.
†Senior Development Engineer.

U = fluid velocity
V = volume
x = quality

Greek Symbols
α = void fraction
β = coefficient of thermal expansion
Γ''' = net rate of vapor generation per unit volume
$(\Gamma'''U)$ = momentum exchange due to vapor generation
μ = viscosity
ν = kinematic viscosity
ρ = density
σ = surface tension
$\underline{\sigma}$ = fluid-fluid stress tensor
$\tau^{\overline{m}}$ = drag force per unit volume
ϕ^2 = two-phase friction multiplier

Subscripts
A = fluid component A
B = fluid component B, bubble
f = fluid
I = interfacial
n = normal
S = interface surface
v = vapor field
vl = between vapor and liquid fields
w = wall
wv = between wall and vapor field

Introduction

Future space missions envision the need for high power levels that are orders of magnitude greater than those of any power source previously launched. These missions will also require advanced propulsion, thermal management, and life support systems. Solar dynamic power cycles have been proposed for the space station. For missions requiring higher power levels, nuclear heat sources will be required. One-phase gas Brayton cycles, alkali metal direct and indirect Rankine cycles, and Stirling power cycles have been proposed. The thermal-hydraulics of a Brayton cycle is the simplest because a one-phase coolant is hardly affected by microgravity conditions and normal gravity fluid-thermal correlations can be used for all but low ranges of coolant flow. Stringent weight, heat transfer, and compactness criteria suggest the use of an alkali metal heat-transfer medium, with a boiling alkali metal direct Rankine system offering advantages over one using a single-phase coolant with an intermediate heat exchanger. Two-phase flow systems also provide weight, heat transfer, and compactness advantages for advanced thermal management and life support systems and for cooling advanced propulsion concepts. However, questions exist regarding the appropriateness of using normal gravity correlations for two-phase pressure drop, heat transfer, and critical heat flux in a microgravity environment (Table 1).

Table 1   Thermal-hydraulic
prediction variables

Flow conditions
Temperature distribution
Pressure distribution
Phase distribution
Heat-transfer-heat flux limits
Turbulence

Activities as part of the U.S. Department of Energy Multimegawatt Space Nuclear Power Program have indicated that an adequate understanding of low gravity flow and heat-transfer behavior does not exist.[1-4] Additionally, the existing reduced-gravity experimental fluid-thermal data base has been found to be insufficient to develop design correlations for two-phase pressure drop, heat-transfer coefficients, and critical heat flux limits for all ranges of considered operating conditions. These correlations are needed for the development and design of space-based nuclear reactors and for other space-based power systems, advanced space-based propulsion systems, and other thermal management and life support systems.

Computer codes represent an important tool for designing complex fluid-thermal systems. Mature computer codes exist that consider single-phase alkali metal and two-phase steam-water flows in normal gravity. To be completely useful for the design and analysis of space-based components, these codes need to be modified to accommodate various operating fluids in variable gravity (Table 2). Therefore, the development of computer codes for the analyses of microgravity systems can be based on existing solution techniques.

Currently available constitutive correlations developed for Earth-based applications are inadequate in reduced gravity. Consequently, the need exists for analytical tools to simulate two-phase flow in a variable-gravity

Table 2   Microgravity technical modeling
concerns

One-phase fluid	Two-phase fluid
Low flow heat transfer	Heat transfer
Low flow pressure drop	Boiling
Turbulence	Condensation
	Heat flux limit
	Pressure drop
	Flow regime
	Interphase effects
	Drag
	Heat transfer
	Mass transfer
	Turbulence
	Freeze-melt

environment. Experimental microgravity two-phase flow data are needed to support the analytical efforts by providing the basis for new constitutive models and correlations for determining flow regime, drag, heat transfer, and pressure drop. Data on alkali metal two-phase forced convection in a normal gravity field is extremely limited; low-gravity data are practically nonexistent. However, low-gravity experiments with two-phase flow of more common fluids (water, air/water, halocarbons) have been more numerous. A comprehensive literature survey of these experiments indicates that both the low-gravity experiments and the acquired data have been limited in various ways. Thus, the usefulness of these past experimental efforts to the design and analysis of space components is marginal, and new experiments will be needed.

The issues associated with low-gravity fluid experimentation are extremely complex. Severe constraints exist on the capability and availability of proper facilities; time and funding levels are incommensurate to the immense task of developing a complete understanding of fluid behavior and heat transfer in a space environment. Given this current situation, the best approach for obtaining low-gravity data would emphasize the phenomena of interest to a component designer and computer code developer.

Analytical Modeling Techniques

Gravity dominates much of the fluid behavior on Earth, especially for two-phase systems. In space, where gravitational forces are small ($<10^{-2} g$), many of the physical forces, which are minor on Earth, exert a significant effect. The effects of the secondary forces are especially felt in two-phase systems. It is instructive to review the conservation equations used as the basis for thermal-hydraulic computer codes to illustrate the effects and changes needed for microgravity considerations.

To obtain meaningful results for modeling low-gravity conditions in any computer code, a number of complex modeling concerns must be addressed. These concerns, which consider the development of analytical models and the required experimental verification, can be divided into two basic categories. The first category addresses items important to one-phase flow conditions, and the second category addresses two-phase concerns.

One-Phase Flow

One-phase flow can be modeled by solving the following conservation equations:

Conservation of mass:

$$\frac{\partial \rho}{\partial t} + \nabla \cdot (\rho U) = 0 \tag{1}$$

Conservation of momentum:

$$\frac{\partial}{\partial t}(\rho U) + \nabla \cdot (\rho U U) = \frac{f p_w}{A} \rho U |U| - \Delta P + \rho g \tag{2}$$

Conservation of energy:

$$\frac{\partial}{\partial t}(\rho h) + \nabla \cdot (\rho h U) = \frac{HA_H}{V}(T_w - T_f) - \nabla \cdot k\nabla T + \frac{\partial P}{\partial t} \qquad (3)$$

For microgravity flow, g must be variable in the conservation equations. Additionally, the correlations for pressure drop and heat transfer must be modified to reflect the loss of significant gravitational forces. For example, at reduced gravity, the absence of buoyancy can affect heat-transfer coefficients for zero or low velocities. Thus, reduced heat-transfer rates are expected because of the lack of buoyancy, and molecular and viscous forces will dominate. Consequently, it is important to include these effects to ensure the accuracy of the predicted flow and temperature fields. In contrast, correlations for high-velocity heat-transfer coefficients for one-phase low-gravity conditions are not expected to differ substantially from those at 1 g. However, the boundary separating the low- and high-flow heat-transfer regimes must be determined.

Although a number of fluid forces that could affect low-gravity fluid-thermal phenomena have been identified, the primary forces are g jitter, residual gravity, drag, and Brownian motion (Table 3).[5,6] The g jitter is the random minor accelerations occurring in a spacecraft caused by components such as offsets from the center of mass, astronaut movement, equipment operation, or small rocket firing. These accelerations amount to random noise at levels less than 10^{-3} g. Residual gravity is present at orbital altitudes. Its magnitude is about 10^{-6} g. These two forces are relatively small for spacecraft considerations. The forces due to working fluid drag (inertial and viscous) and Brownian motion tend to determine low-gravity fluid-thermal behavior.

In microgravity, the fluid-thermal behavior of a one-phase fluid is dominated by convection (viscous or inertial) and diffusion (mass and thermal). As previously stated, if the flow velocity is large enough, the inertial drag will overwhelm the other forces. At low velocity, viscous convection and diffusion effects will dominate. As on the Earth, the Reynolds number, which compares inertial to viscous forces should be indicative of the type of flow present. Similarly well-known nondimensional numbers, such as the Nusselt (convection/conduction), Prandtl (momentum diffusion/heat diffusion),

Table 3   One-phase microgravity fluid-thermal forces

g jitter $(<10^{-3}\,g)$
Residual gravity $(<10^{-6}\,g)$
Drag forces
Inertial convection
Viscous convection
Thermal gradients
Concentration gradients
Brownian motion
Thermal diffusion
Mass diffusion

Table 4   Two-phase microgravity fluid-thermal forces

g jitter ($<10^{-3}g$)
Residual gravity ($<10^{-6}g$)
Drag forces
Inertial convection
Viscous convection
Thermal gradients
Concentration gradients
Surface tension effects: liquid-gas and fluid-surface interactions
Marangoni flow: interfacial force from low to high surface tension
Brownian motion
Thermal diffusion
Mass diffusion

Grashof (gravitational force/viscous force), and Froude (inertial force/gravitational force) numbers, would still be appropriate to describe fluid-thermal conditions in reduced gravity.

Two-Phase Flow

Two-phase flow systems are subject to the same four basic forces as one-phase flow. However, because two phases are present, additional force components must be considered. Thus, Marangoni flow and surface tension forces (Table 4) become important drag components, as does interfacial diffusion. To relate the importance of the various forces, it is convenient to use nondimensional numbers. In addition to the nondimensional relations previously mentioned for one-phase applications, the Bond number (gravitational/surface tension) and the Weber number (surface tension/inertial force) are also important for describing two-phase flow (Table 5).

It is appropriate to describe Marangoni flow because this type of flow is negligible compared to buoyancy driven convection and is therefore not well known to individuals familiar with normal gravity flow. Marangoni flows can exist where there is a free surface between two phases (Figs. 1 and 2).[5] The variation of surface tension due to thermal or concentration gradients along a gas-liquid interface sets up surface tangential stresses that give rise to an interfacial force in the direction of increasing surface tension.

Table 5   Important dimensionless parameters

Reynolds number	$[\rho(UL)/\mu]$	Inertial force/viscous force
Nusselt number	$[(HL)/k]$	Heat convection/heat conduction
Prandtl number	$[(C_p\mu)/k]$	Momentum diffusion/heat diffusion
Grashof number	$[(\rho^2 g\beta\Delta T L^3)/\mu^2]$	Gravitational force/viscous force
Froude number	$[U^2/(gL)]$	Inertial force/gravitational force
Weber number	$[(\rho U^2 d)/\sigma]$	Inertial force/surface tension
Bond number	(We/Fr)	Gravitational force/surface tension
Schmidt number	(ν/D_{AB})	Momentum diffusion/mass diffusion

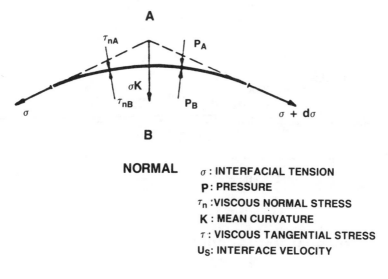

NORMAL σ : INTERFACIAL TENSION
         P : PRESSURE
         τ_n : VISCOUS NORMAL STRESS
         K : MEAN CURVATURE
         τ : VISCOUS TANGENTIAL STRESS
         U_S : INTERFACE VELOCITY

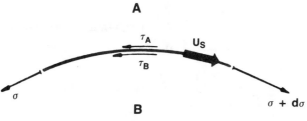

TANGENTIAL

Fig. 1   Stress diagram at a free surface between two fluids. (If forces are unbalanced, motion from low tension to high tension areas will occur along the surface.[5])

The moving surface molecules drag the adjacent fluid, causing flow from the region of lower tension to the region of higher surface tension. For gas bubbles such liquid flow can result in bubble motion. Thus, for a bubble present in a temperature gradient, the motion of the surrounding liquid would cause a jet that would move the bubble toward the hotter region unless opposed by other forces (Fig. 3).[7] (Surface tension decreases with temperature increase.) The importance of Marangoni flow in a low-gravity environment must be determined because this effect can result in serious consequences for boiling heat transfer and critical heat flux. The motion of a bubble toward a hot surface would encourage vapor blanketing, degrade heat transfer, and hasten the transition to film boiling.

Marangoni flow has been observed in microgravity under certain conditions. A 4-min rocket provided a microgravity environment for an experiment on bubble migration in molten glass ("fining"), as reported by Wilcox et al.[8] A platinum heating strip melted a sample of sodium borate

FORCES

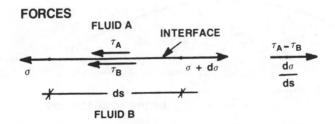

THERMAL MARANGONI FLOW

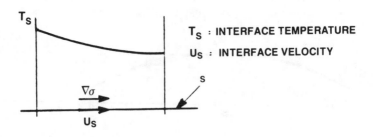

VELOCITY FIELD

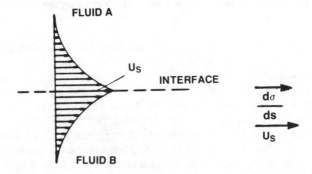

Fig. 2   Marangoni thermal convection. (A thermal gradient along the surface will set it in motion. The velocity will be highest at the surface, and internal fluid will be dragged along.[5])

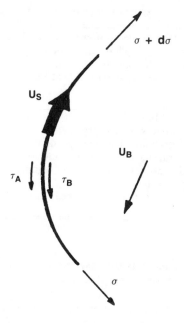

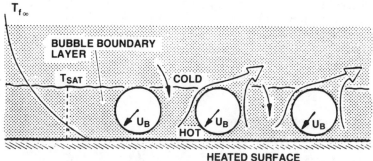

Fig. 3   Thermocapillary mechanism of subcooled boiling.[7]

glass, which contained entrapped voids. During the low-gravity portion of
the flight, distinct migration of the bubbles toward the hotter portion of the
sample was noted. This observation is in agreement with the Brown model[7]
of thermocapillary bubble migration, which predicts bubble motion against
a thermal gradient.

The analysis of two-phase flow can be accomplished using a homogeneous mixture assumption, or more complicated methods, such as a
multifield or drift flux approach (Table 6). All treatments require the
presence of g in the conservation equations in a manner similar to the
one-phase approach. The two-phase techniques must address the one-phase
concerns plus additional two-phase items. The homogeneous mixture approach represents the simplest of the two-phase methods. The principal

Table 6   Required two-phase modeling relations

Homogeneous approach variables
Pressure drop multiplier-wall friction
Wall heat transfer-boiling and condensation
Boiling heat flux limits-critical heat flux (CHF)
Turbulence
Multifield approach variables
Wall friction-flow regime dependent
Interfacial drag-flow regime dependent
Wall heat transfer-boiling and condensation
Interfacial heat/mass transfer-flow regime dependent
Flow regime map
Boiling heat flux limits-critical heat flux (CHF)
Turbulence

limitation to a homogeneous two-phase technique is the inherent simplicity of the mixture representations that does not permit the calculation of phase slip and nonequilibrium. A brief look at the governing equations is instructive. The three equations defining conservation of mass, momentum, and energy for a homogeneous two-phase mixture are similar to the one-phase equations and have several terms that are directly or indirectly dependent on gravity.

Conservation of mass:

$$\frac{\partial \rho}{\partial t} + \nabla \cdot (\rho U) = 0 \tag{4}$$

Conservation of momentum:

$$\frac{\partial}{\partial t} (\rho U) + \nabla \cdot (\rho U U) = \frac{f\phi^2 P_w}{A} \rho U |U| - \Delta P + \rho g \tag{5}$$

Conservation of energy:

$$\frac{\partial}{\partial t} (\rho h) + \nabla \cdot (\rho h U) = \frac{H A_H}{V} (T_w - T_f) - \nabla \cdot k \nabla T + \frac{\partial P}{\partial t} \tag{6}$$

In Eq. (5), the momentum equation, the magnitudes of f and ϕ^2 are gravity-dependent, and the body force is explicitly a function of gravity. In Eq. (6) H can be gravity-dependent. For homogeneous two-phase flow, setting g equal to zero and inputting correlations or values for f, ϕ^2, and H that correspond to a low-gravity environment is one approach for modeling low-gravity effects. The major effort then lies in experimental determination of f, ϕ^2, and H. Limited two-phase microgravity experiments indicate that two-phase pressure drops are greater than those on the Earth. Also, because the lack of buoyancy at low gravity results in the increased

Table 7 Required two-phase variable-gravity experimental correlations

Pressure drop correlation
Interfacial drag
Heat-transfer coefficient
Heat flux limits-critical heat flux (CHF)
Flow regime determination
Interfacial heat-transfer coefficient
Interfacial mass transfer coefficient
Turbulence measurements

importance of other heat-transfer mechanisms, existing two-phase heat-transfer correlations are inadequate. Therefore, low-gravity single- and two-phase heat-transfer coefficient correlations and two-phase pressure drop multiplier relations must be experimentally determined and analytically described.

The use of more complicated two-phase relations for a low-gravity environment represents advantages over simpler approaches and is in fact required for analyses where phase separation, slip, and thermal nonequilibrium conditions exist. However, these approaches do introduce additional complications. The basic conservation equations must be written to introduce a variable g term, and the low-gravity correlations for two-phase wall friction factor and wall heat transfer must be appropriately incorporated. However, additional information must be developed to determine flow regime and interphase drag and heat transfer (Table 7).

Again, a look at the governing conservation equations yields some insight. For a multifield approach, a minimum of three equations for each phase are required to define mass, momentum, and energy conservation. These equations have terms that are directly or indirectly dependent on gravity. Compared to the single-phase and homogeneous two-phase models, more relationships for unknown terms are required. An equation set for the vapor component of a 6-equation system that individually models the liquid and gaseous phases follows. Similar considerations exist for 9-equation systems that model the liquid, droplet, and gas fields and 12-equation systems that model the liquid, droplet, gas, and bubble fields.

Conservation of mass:

$$\frac{\partial}{\partial t}(\alpha \rho_v) + \nabla \cdot (\alpha \rho_v U_v) = \Gamma''' \tag{7}$$

Conservation of momentum:

$$\frac{\partial}{\partial t}(\alpha \rho_v U_v) + \nabla \cdot (\alpha \rho_v U_v U_v) = -\alpha \Delta P + \rho_v g$$

$$+ \nabla \cdot (\alpha \underline{\sigma}_v) + \tau_{wv}''' - \tau_{I_{vl}}''' + (\Gamma''' U) \tag{8}$$

Conservation of energy:

$$\frac{\partial}{\partial t}(\alpha \rho_v h_v) + \nabla \cdot (\alpha \rho_v h_v U_v) = -\nabla \cdot (\alpha g_v)$$

$$+ \Gamma''' h_g + q'''_{I_v} + Q'''_{wv} + \alpha \frac{\partial P}{\partial t} \qquad (9)$$

The momentum equation, Eq. (8), requires 0-g correlations for the interfacial and wall drag terms $\tau'''_{I_{vl}}$ and τ'''_{wv}. Furthermore, the interphase mass and heat-transfer rates per unit volume in all equations are gravity-dependent, and a low-gravity correlation must be developed. Also, all terms containing g must be set to a variable value. The energy equation, Eq. (9), requires additional relationships for the determination of heat transfer between the phases and at the solid boundaries. All of these heat- and mass-transfer relations are dependent on the flow regime of the fluid mixture. Thus, two-phase flow regime maps for low and variable gravity must be determined by experiment and described analytically. These relations must reflect the increased importance of surface tension at liquid-gas interfaces and molecular interaction of the liquid at solid surfaces to permit an adequate flow regime determination. These items would also be important in the development of a liquid free-surface calculational ability.

Computer Programs

Survey, Characteristics, and Assessment

A survey of several major computer codes has been performed to determine their availability and appropriateness for the analysis of proposed space thermal-fluid components under microgravity conditions.[1-3] The survey considers existing multidimensional single- and multiphase computer codes using details provided from the code manuals or other supporting information. A ranking system was developed to assign relative importance to existing code capabilities in various categories. The considered codes represent the state-of-the-art in computer modeling, especially for multiphase flow. The considered computer codes have been divided into three categories:

1) System computer codes: This group of codes is intended to analyze flow conditions in fluid-thermal components and in external piping. These codes are generally simpler than the complex multidimensional codes and thus should be used for preliminary design and assessment.

2) Multidimensional one-phase computer codes: The computer codes developed for this category could be used for exact component assessments and analyses.

3) Multidimensional two-phase computer codes: This group of codes is generally used for performing more accurate component assessments and analyses.

To obtain meaningful results from any of the considered codes, a number of modifications must be made for microgravity applications. Depending on the code, the extent and complexity of the modifications may vary. The developed rating system reflects the desired code characteristics. The recommended codes are those that possess the highest cumulative ratings obtained by using the weighting system.

The code characteristics and weightings are representative of the computer code requirements. The recommended codes must be well documented, and the coding must be structured and understandable. The input must be straightforward, and the output should provide all of the required information. Graphics capabilities are not considered as important. Stable steady-state and transient computational capabilities are considered important concerns, as is the ease of adding desired fluid thermophysical properties.

The presence of certain computational models has affected the weightings. Some codes already possess heat-transfer and pressure drop capabilities and logic for calculating both one- and two-phase conditions. The presence of these calculational abilities is reflected in the relative weightings. The ease of eliminating the gravitational effect from the conservation equations is also important. Although surface tension effects will be important in low gravity, few of the currently available computer models consider surface tension effects explicitly.

Other considerations arise from the current user's perception of the code. Acceptance and use of a code are indicative of the stability of the solution procedure and of the code's usability. Documentation is an important item. The lack of a user's manual severely limits the use of a code and essentially results in elimination from consideration. The lack of a programming manual is less important as long as the code's language, structure, and logic are understandable, or the author's support for the code is available.

Code verification is ultimately required and needs to be performed by comparing code predictions to exact analytical solutions and experimental data. Because code modifications will be necessary for the analysis of proposed components in a low-gravity environment, experimental verification is a necessity.

One-Dimensional System Codes
The nature of thermal-hydraulics power, thermal management and propulsion systems suggests the use of a comprehensive computer modeling approach that considers the entire flow loop. Important feedback effects are provided by the coupling of all of the system component models, such as the evaporators, pumps, and radiators, in addition to the valves and piping. It is appropriate to identify systems codes that would provide a total package to analyze the simple and complex portions of the system. The systems portion of the code could supply the appropriate boundary conditions to the component models to calculate the steady-state and transient conditions.

The need for a coupled systems code is based on experience obtained in the analysis of nuclear power systems. An early approach for simulating a severe loss of coolant accident in a pressurized water reactor with a detailed multidimensional reactor computer model attempted to use transient boundary conditions obtained from a previously run systems code. Differences between the core and system calculations quickly accumulated, resulting in a mismatch between the applied boundary conditions and the computed reactor core conditions. None of these simulations was successful until the system calculations were coupled to the reactor core simulation.

Generally, system codes are one-dimensional and are capable of analyzing the various components present in a thermal-fluid system. These components include evaporators, condensers, reactor cores, heat exchangers, pumps, turbines, radiators, valves, and pipes. Some system codes also include three-dimensional models. Many of these models are simplified versions of more complex models.

A system code is desired that could calculate operational steady-state conditions and transient conditions that could result in one- or two-phase flow. The desired system code characteristics and capabilities for determining variable-gravity conditions include the following:

1) numerical stability;

2) compressible flow capabilities;

3) two-phase flow models (the solution of separate conservation equations for the liquid and gaseous fields is desired for modeling variable-gravity conditions; as a minimum, a two-field model is desired);

4) turbulence models (at present no system code contains turbulence models; their addition will be important for modeling low-gravity conditions);

5) normal-gravity one-phase heat-transfer models;

6) normal-gravity two-phase heat-transfer models including condensing, boiling, and critical heat flux (variable-gravity correlations must be added when available);

7) two-phase pressure drop capabilities (variable-gravity models must be added when available);

8) radiation heat-transfer capabilities;

9) solid heat-conduction modeling; and

10) transient and steady-state calculational ability.

It is recognized that the thermal-fluid correlations currently included in all systems codes would not necessarily be appropriate for low-gravity conditions. However, their presence would provide a basis for model development since the position in the numerical solution procedure is already defined.

Using these considerations, a number of system codes were evaluated. Generally, the most promising codes possessed a multifield two-phase flow modeling capability. As stated earlier, this approach is considered the best approach for analyzing low-gravity two-phase flow because the two phases, or fields of the two phases (liquid, drops, vapor, bubbles), are separately solved in a set of conservation equations that include interfacial heat and mass transfer effects. The FLUINT[9] computer code does not possess the desired method for predicting low-gravity two-phase flow conditions. FLUINT's homogeneous two-phase modeling approach is not capable of predicting the unequal velocity and temperature conditions expected in low gravity. It would also be difficult to empirically correlate the separated-flow phenomena in low-gravity condition with the homogeneous conservation equations.

Promising codes that could be used for low-gravity fluid-thermal calculations include ATHENA,[10-12] COBRA/TRAC,[13] FLOW-NET,[14,15] and TRAC-BD1.[16,17] All of these codes are based on methodologies developed as part of the light water reactor (LWR) programs. Only the ATHENA code has been expanded to include space-related concerns such as preliminary variable-gravity effects, heat pipe components, and a library of fluid properties including alkali metal and cryogenic fluids. The ATHENA, COBRA/TRAC, and TRAC-BD1 codes solve the conservation equations using nonequilibrium, two-fluid models with a minimum of two fields. The FLOW-NET code employs a slip flow method for the solution of the conservation equations. It solves continuity equations for the two phases, a mixture momentum equation with a slip relation, and a mixture energy equation. The presence of a thermal radiation model is also considered important. The ATHENA and TRAC-BD1 codes are the only codes that possess participating media thermal radiation models. Summaries of the characteristics of all these codes are contained in the following section and are summarized in Table 8. The FLUINT computer program is described in Chapter 2. This listing represents the technically promising codes currently available as determined by the assessments performed in Refs. 1–3.

System Computer Code Summaries

ATHENA.[10-12] The Advanced Thermal-Hydraulic Energy Network Analyzer (ATHENA) code was developed at the Idaho National Engineering Laboratory to perform transient or steady-state simulation of advanced thermal-hydraulic systems such as space reactors, fusion devices, or other thermal management systems. The code is an outgrowth of the RELAP5/MOD2 code. ATHENA solves one-dimensional transient flow equations for two-fluid, two-phase, multicomponent flow. Properties for a number of operating fluids, including alkali metal and cryogenic fluids, are included as polynomial curve fits. Noncondensable gases are treated as ideal gases. Other fluids may be included by specifying the appropriate polynomial constants. Although the code is written to solve two-field conservation equations for nonthermal equilibrium, several options are available for simpler flow models such as homogeneous flow and thermal equilibrium models. The system model is solved numerically using a semi-implicit finite-difference technique. The ill-posed nature of the numerical solution is rendered well-posed by the introduction of artificial viscosity terms. The semi-implicit scheme uses a direct sparse matrix solution technique for time-step advancement. The code was written in FORTRAN V for Cyber and Cray computers.

The code includes several component models such as pumps, valves, one- and two-dimensional heat-conducting structures, turbines, heat pipes, control system logic, and nuclear reactor kinetics. Special models are also

Table 8 System computer code comparisons

	ATHENA	COBRA/TRAC	FLUINT	FLOW-NET	TRAC-BD1
Transient or steady state	Both	Transient	Both	Transient	Both
Fluid properties	Input	Code mod	Input	Input	Code mod
Gravity variance	Input	Code mod	Input	Input	Code mod
Compressible flow	Yes	Yes	No	Yes	No
Flow model	Two-fluid	Two-fluid Drift flux	Homogeneous	Two-fluid slip	Two-fluid
Phase temperatures	Unequal	Unequal	Equal	Equal	Unequal
Phase velocities	Unequal	Unequal-drift flux	Equal	Unequal-slip model	Unequal
1-g Boiling	Yes	Yes	Yes	Yes	Yes
1-g Boiling limit (CHF)	Yes	Yes	No	No	Yes
1-g Condensing	Yes	Yes	Yes	Yes	Yes
1-g Two-phase pressure drop	Wall/interface drag	Wall/interface drag	Wall	Martinelli/interface drag	Wall/interface drag
1-g Flow regime	Yes	Yes	No	No	Yes
Flow choking	Yes	Yes	No	No	Yes
Turbulence	No	Yes	No	No	No
3-D Flow component	No	Yes	No	No	Yes
Thermal conduction	2-D	2-D	Lumped parameter		2-D
Thermal radiation	Yes	No	Yes	No	Yes
Component models	Pipes Pumps Valves Turbines Heat pipes MHD effects Control logic	Pipes Pumps Valves	Pipes Pumps Valves HX	Pipes Valves	Pipes Pumps Valves Turbines Separators HX

included for describing choked flow, tees, area changes, and magnetohydro-dynamic (MHD) effects. One- or two-dimensional heat-conduction solids may be specified. The code possesses a full participating media thermal radiation model. The gravitational term in the conservation equation is an input to permit modeling of variable-gravity conditions.

COBRA/TRAC.[13] The COBRA/TRAC computer code, which was devel-oped at Battelle's Pacific Northwest Laboratories, is a combination of the three-dimensional COBRA-TF code and the TRAC-PD2 systems computer program. The code was written in FORTRAN and is operational on CDC and Cray computers. COBRA-TF provides a two-fluid, three-field repre-sentation of compressible, two-phase flow in a three-dimensional geometry such as a reactor vessel core. The three fields are continuous vapor, continuous liquid, and entrained liquid drops. The conservation equations and the heat transfer to and within one- or two-dimensional solid heat-conducting structures are solved using a semi-implicit, finite-difference numerical technique on an Eulerian mesh. The three-dimensional formula-tion may be represented with rectangular Cartesian coordinates or using a subchannel approach.

TRAC-PD2 is a systems code designed to model the behavior of an entire fluid-thermal system. The code employs a one-dimensional five-equation drift flux representation of two-phase flow. The special compo-nents system models include pumps, valves, pressurizers, steam generators, tees, and accumulators. Thermodynamic properties are obtained from data tables or polynomial equations. One- or two-dimensional heat-conducting structures can be specified.

FLOW-NET.[14,15] The FLOW-NET computer code was written to solve one-dimensional nonequilibrium, two-phase flow network transients using a drift flux method. The code solves continuity equations for the mixture, vapor, and gas. The momentum and energy equations are solved using mixture equations only. The momentum equation accounts for slip and for interfacial mass and momentum transfer. The thermodynamic properties for any two-phase working fluid are input as curve fit relations. The inert gas is treated as an ideal gas. Internal numerical methods permit the use of an automatic time-step control, a one-dimensional mesh generator, and implicit or explicit solution techniques. The field, constitutive, and state equations are solved using the implicit continuous-fluid Eulerian (ICE) finite-difference method, which is applicable to flow at all speeds.

TRAC-BD1.[16,17] TRAC-BD1, developed at the Idaho National Engineer-ing Laboratory, is a fully nonhomogeneous, nonequilibrium, two-component, two-fluid thermodynamic systems computer code. It is in-tended to predict steady-state and transient two-phase flow in a boiling water reactor (BWR). It was written in FORTRAN 77 for a Cyber computer. The code includes a three-dimensional thermal-hydraulic model-ing capability. Semi-implicit finite-difference techniques are used for both the one- and three-dimensional components. The one-dimensional compo-nents have a numerical solution capable of violating the material Courant limit. Components that may be specified include turbines, heat exchang-ers, phase separators, pumps, and valves. Two-dimensional cylindrical heat-conduction structures may also be specified. The code possesses a full participating media thermal radiation model. Thermophysical properties for water are based on polynomial curve fits. The noncondensable gas is assumed to be ideal.

Multidimensional Codes

Three-dimensional computer codes are needed to perform detailed design analyses of proposed space-based fluid-thermal components. These detailed models would also verify the adequacy of the simpler component models used in the systems codes. Specific components that may need the detailed analyses include phase separators, heat exchangers, evaporators, and condensers.

To verify the adequacy of the detailed three-dimensional analyses, experimental verification is needed. It may not be possible to perform experiments on a component over an entire range of expected normal and transient operating conditions. It may also be impossible to test a component in the exact configuration and for the exact conditions under which it will ultimately operate. Thus, mathematical modeling using the detailed three-dimensional computer codes is necessary to expand the performance data base over the range of steady-state and transient operating conditions.

A number of three-dimensional computer codes are currently available for use in a normal-gravity environment. None of these codes has undergone extensive verification for application to variable-gravity conditions. The thermal-hydraulic concerns for modeling low-gravity conditions are independent of the number of dimensions being considered in an analysis. Of course, a three-dimensional model introduces significant complexity over the simpler one-dimensional models.

It is generally expected that the currently available computer codes could be modified to address variable-gravity conditions. The FLOW3D[18,19] and NASA-VOF[20,21] computer codes are examples of this process. Earlier versions of these codes were originally developed and used for applications in the nuclear industry. These codes represent the best available methods of analyzing tank slosh problems in variable gravity. The codes possess a simple method of predicting surface movements between the liquid and gas volumes in a tank. However, not all the fluid-thermal interactions are modeled. Both codes solve the Navier-Stokes equations for the liquid only. Specifically, continuity and momentum relations for the liquid are satisfied, but the liquid energy equation is not addressed. Additionally, conservation equations for the gas are not present. More detailed codes are needed to predict the fluid-gas behavior in a tank containing fluids, such as cryogenics, where wall heat transfer, interfacial heat and mass transfer, and vapor conservation relations are important.

A number of available computer codes can serve as the basis for developing low-gravity multidimensional computer models. These codes can be divided into those addressing one- or two-phase conditions. The one-phase codes may need little modification for analyzing low-gravity conditions. However, the two-phase codes will require extensive modification in addition to experimental verification. As previously stated, the multifield method, which separately solves the conservation equations for different fields (liquid, vapor, drops, bubbles), is the most attractive method of predicting two-phase behavior. A summary of available one- and two-phase multidimensional codes is given in Table 9. This listing represents the technically promising computer codes currently available, as determined by the assessments prepared for the Multimegawatt Space Power Program.[1-3]

Mass- and Heat-Transfer Experimental Data and Trends

In addition to the obvious changes in the conservation equations necessitated by the variance of gravity, the fluid-thermal relations used in the conservation equations will be affected by the absence of gravity. These relations were mentioned in the previous section. The following sections more fully discuss these items. Table 10 summaries the available low-gravity experimental data, and Table 11 presents the known experimental trends.

The scope and extent of experiments in two-phase low-gravity flows have been determined largely by the methods and facilities. The readily available means (e.g., drop tower, aircraft flying parabolic trajectories) have seen the greatest use; closer approximations to a long-term microgravity environment (e.g., rocket, Space Shuttle) have been employed much less frequently. The restrictions imposed by the minimum attainable gravity field and its duration necessarily delimit data in various ways. Additional constraints on the type of fluid, size of the experimental package, power available to run the experiment, and instrumentation and data acquisition equipment have strongly influenced the nature of the data. As a result, the operable mechanisms in reduced-gravity two-phase convection have not been elucidated, nor have quantitative heat-transfer and hydrodynamic correlations been developed. Nevertheless, past experiments have yielded some insight into the phenomena dominating fluids as the gravity effect is removed.

Flow Regime Determination

This relation is primarily required for the solution of the multifield correlation conservation equations. The homogeneous relations do not directly employ flow regime considerations; however, they can be used in an indirect form in some fluid-thermal relations. Because reduced-gravity heat transfer and pressure drop will be strongly affected by flow regime, it is felt that the use of the multifield equations would more easily model variable-gravity conditions. Similarly, flow regime determination is probably the most important microgravity two-phase phenomenon because it affects both flow and heat-transfer conditions.

A fairly recent experiment delved into the topics of flow regime and pressure drop.[31] Two-phase flow of air and water in a circular channel was examined first on Earth, then in an aircraft simulating $0\,g$ for about 20 s per trajectory. Analysis indicated a downward shift of regime boundaries at reduced gravity. That is, at a given quality, the transition from distributed (bubbly or mist) to segregated plus intermittent (slug) to segregated

Table 9 Multidimensional one-phase computer code comparisons

	COBRA-WC	COBRA-SFS	COMMIX-1B	FLUENT	PHOENICS	NASA-OF FLOW V3D	TEMPEST
Reference	22	23	24	25	26	18, 19, 20, 21	27
Transient or steady state	Both	Both	Both	Trans	Both	Trans	Both
Dimensions	3-D	3-D	3-D	3-D	3-D	3-D	3-D
Fluid properties	Input	Input	Input	Input	Input[a]	Input	Input
Gravity variance	Code mod	Code mod	Input	Code mod	Input[a]	Input	Input
Compressible flow	No	No	Yes	No	Yes	No	No
Mass, momentum equations	Yes	Yes	Yes	Yes	Yes	Yes	Yes
Energy equation	Yes	Yes	Yes	Yes	Yes	No	Yes
Turbulence	Yes	Yes	Yes	Yes	Yes	No	Yes
Heat transfer	Yes	Yes	Yes	Yes	Yes	No	Yes
Thermal conduction	1-D	3-D	1-D	3-D	3-D	None	3-D
Thermal radiation	No	Yes	No	No	Yes	No	Yes
Free surface calculation	No	No	No	No	No	Yes	No

(Table 9 continued on next page)

Table 9 (cont.) Multidimensional two-phase computer code comparisons

	COBRA-NC	COBRA/TRAC	COMMIX-2	PHOENICS	TRAC-PF1	TRAC-BD1
Reference	28	13	29	26	30	16,17
Transient or steady state	Trans	Trans	Trans	Both	Both	Both
Dimensions	3-D	3-D	3-D	3-D	3-D	3-D
Fluid properties	Code mod	Code mod	Code mod	Input[a]	Code mod	Code mod
Gravity variance	Code mod	Code mod	Code mod	Input[a]	Code mod	Code mod
Compressible flow	Yes	Yes	No	Yes	Yes	Yes
Flow model	Two-fluid 3-field	Two-fluid 3-field	Two-fluid 5-equation	Input[a]	Two-fluid 8-equation	Two-fluid 8-equation
Noncondensables	Yes	No	No	Input[a]	Yes	Yes
Phase temperatures	Unequal	Unequal	Equal	Input[a]	Unequal	Unequal
Phase velocities	Unequal	Unequal	Unequal	Input[a]	Unequal	Unequal
1-g Boiling	Yes	Yes	Yes	Input[a]	Yes	Yes
1-g Boiling limits (CHF)	Yes	Yes	Yes	Input[a]	Yes	Yes
1-g Condensing	Yes	Yes	Yes	Input[a]	Yes	Yes
1-g Two-phase pressure drop	Yes	Yes	Yes	Input[a]	Yes	Yes
Turbulence	Yes	Yes	No	Input[a]	No	No
Thermal conduction	2-D	2-D	1-D	3-D	2-D	2-D
Thermal radiation	No	No	No	Yes	No	Yes

[a] User specified input coding.

Table 10   Reduced-gravity two-phase experiments summary

Type	Investigated	Fluids	Flow	Heat transfer	Liquid state	Conclusions	Comment	Reference
Drop test	Single bubble growth	Water, toluene, hexane	No	Not examined	Saturated	Simple correlations found for growth and shape variance with g	Not directly useful for present needs	Cooper et al.[42]
Drop test	Vapor generation at liquid surface and in bulk fluid	R-11	No	Not examined	Saturated	No vapor generated (i.e., bubbles) in bulk liquid upon venting at 0 g	Not relevant	Labus et al.[43]
Drop test	Forced convection boiling at low heat flux and velocity	Distilled water	Yes	Not examined	Slightly subcooled	Simple correlation found for bubble diameter versus evaporation layer	Not relevant	Cochran[44]
Drop test	Pool boiling; effects of surface tension, viscosity, and subcooling	Water, ethanol-water, sucrose-water	No	Varied in some tests	Various subcoolings	Water boiling (i.e., bubble size) independent of g, at high subcooling	Trends of interest	Cochran et al.[45]
Drop test	Pool boiling (nucleate plus film)	Various	Yes	Yes	Various	Critical flux proportional to $g^{\frac{1}{4}}$	Preliminary results; g fields high (~0.01)	Siegel[37]
Drop test	Pool boiling	N_2, R-113	No	Yes	Saturated	Surface superheat less in 0 g	Transient results; g fields high (~0.004)	Oker and Merte[46]

(Table 10 continued on next page.)

Table 10 (cont.) Reduced-gravity two-phase experiments summary

Type	Investigated	Fluids	Flow	Heat transfer	Liquid state	Conclusions	Comment	Reference
Aircraft	Flow regimes and pressure drop	Air, water	Yes	None	N/A	g Level influences flow regime and thus pressure drop	Good basis for future work	Heppner et al.[31]
Aircraft	Flow regimes in condensation	R-12	Yes	Qualitative	Superheated/saturated	Baker chart valid; no reduction in heat transfer with 0 g	Not directly useful	Williams[32] Keshock et al.[33] Williams et al.[34]
Aircraft	Pressure drop and phase velocities in condensation	Hg	Yes	Yes	Superheated/dropwise condensation	Little difference between 0- and 1-g data	Not directly useful	Albers and Macosko[35] Namkoong et al.[36]
Rocket	Void (i.e., bubble) migration in molten glass	Sodium borate glass	No	Not examined	N/A	Bubbles migrate against thermal gradient at 0 g	Not directly useful	Wilcox et al.[8]
Drop test, aircraft	Flow regime	Water, N_2	Yes	Not examined	Subcooled	Flow regime data and model	Useful	Dukler et al.[66]
Aircraft	Flow regime	Water, N_2	Yes	Not examined	Subcooled	Flow regime data and model	Useful	Lee et al.[60] Lee and Best[61]
Aircraft	Boiling, condensing	Water	Yes	Yes	Various	Comparisons with normal-gravity conditions	Useful	Kachnik et al.[59] Cuta and Krotiuk[65] Cuta et al.[64]
Aircraft	Flow regime, pressure drop	Freon (R-114)	Yes	No	Saturated	Flow regime and pressure drop data	Useful	Hill et al.[74,75]
Aircraft	Boiling, condensing	Freon (R-114)	Yes	Yes	Saturated	Swirl flow evaporator, RFMD, condensing equipment evaluation	Useful	Hill et al.[74,75]
Drop tower	Pool boiling	LN_2, Hg	No	Yes	Saturated	Pool boiling at reduced and high gravity	Useful	Merte[47]

Table 11 Microgravity trends compared to 1-g conditions

		One-phase	Two-phase
Low flow (<140 kg/m²/s)	Pressure drop	Comparative	Comparative or higher
	Heat-transfer coefficient[a]	Lower	Boiling: comparative but higher; condensation: comparative
	Flow regime	—	Shifted
	Critical heat flux	—	Pool boiling: varies $g^{1/4}$; Flow boiling: lower?
High flow (>230 kg/m²/s)	Pressure drop	Comparative	Comparative but higher for equivalent flow regime
	Heat-transfer coefficient[a]	Comparative	Boiling: comparative but lower; condensation: comparative but lower
	Flow regime	—	Shifted
	Critical heat flux	—	Lower?

[a]Nonmetallic fluids.

(annular) flow occurred at a lower total mass flow rate in reduced gravity. Initial testing confirmed this trend, but not its magnitude. However, a repeat test nearly agreed with the analytical predictions. The low-gravity pressure drop was significantly higher than that for 1 g and was ascribed to the change in flow regime, which was a consequence of increased turbulence in 0 g.

A similar experimental effort was devoted to condensation of Freon-12.[32-34] Although the test section was well instrumented, only qualitative results are reported. From photographic records, it appears that the flow regimes observed conform reasonably well to Baker flow regime chart predictions. The flow at 0 g was notably less irregular than that at 1 g, which is somewhat at variance with the trend noted earlier. Because there was little discernible difference in condensation lengths, it was also hypothesized that heat transfer was unaffected by g level.

Some preliminary studies used mercury in reduced-gravity condensation experiments. Albers and Macosko[35] reported practically the same pressure losses at normal and 0 g in a constant-diameter tube. Both losses were greater than predicted by the Lockhart-Martinelli correlation at low vapor qualities. But in the high-quality region of the condensing tube, the pressure drop from the Lockhart-Martinelli correlation was within $\pm 70\%$ of the measured pressure loss. It was believed that fog-flow theory, which postulates re-entrainment of condensed droplets back into the vapor stream, best explained the data since visualization was not possible in the stainless steel test section.

A photographic experiment conducted by Namkoong et al.[36] with mercury vapor condensing in glass tubes failed to support the fog-flow hypothesis. In tubes with diameters $\geqslant 1$ cm, the distribution of drops at the wall was concentrated on the tube bottom in 1-g conditions. Zero-gravity conditions led to a uniform distribution of droplets, both in the droplet stream and at the wall.

Several investigators are currently working to experimentally determine microgravity flow regime maps and to develop relations to understand the mechanisms of transition between different flow patterns using data from drop tower and parabolic aircraft experiments. These investigations are discussed more fully in the section on Recent Experimental Efforts.

The investigators have observed three low-gravity flow regimes, the dispersed (bubbly or mist), slug, and annular flow regimes. The appropriateness of the recommended correlations for determining low-gravity flow regimes are subject to further verification and model improvement. However, these relations represent an important first step in understanding the forces that define low-gravity flow regimes.

Pressure Drop

As can be expected, two-phase pressure drop is also dependent on flow regime. Thus, the use of a two-phase pressure drop multiplier (ϕ^2) does not present the most appropriate means of calculating frictional losses. The multifield approach affords a better means of calculating pressure drop but results in calculational complications. Pressure drop for multifield flow must account for wall friction and interfacial drag. Thus, the multifield approach must supply wall and interfacial drag correlations dependent on flow regime and other flow conditions such as slip.

No correlations currently exist that account for low-gravity conditions in the calculation of two-phase pressure drop. Past researchers have indicated qualitative trends but no quantitative data. Their findings indicated that low-gravity two-phase pressure drops are greater than those obtained from equivalent normal-gravity two-phase flow conditions.[31] Recent experiments addressing low-gravity pressure drop are discussed in the section on Recent Experimental Efforts.

Heat Transfer

An extensive literature search has revealed that low-gravity heat transfer has been studied for a number of years. These activities have not adequately addressed most thermal-hydraulic issues, but serve as a basis for further work. A review article by Siegel[37] discusses and summarizes pre-1967 reduced-gravity experimental data. Another review article[38] recounts the history of early low-gravity materials processing experiments, some of which investigated heat flow and convection. These efforts consisted typically of fairly simple experiments that measured only a few parameters. Oran[39] provides an overview of the current materials experiments before the Challenger accident, and Carruthers[40] supplies the rationale motivating the experiments. Another paper[41] reviews some of the same experimental work as discussed here and arrives at similar conclusions.

Low-gravity heat-transfer concerns can be divided into four areas of interest: one-phase, boiling, condensation, and critical heat flux. A discussion of each area of interest follows.

One-Phase

As previously stated, low-gravity, one-phase, forced convection, heat-transfer mechanisms are not expected to differ from those observed in normal gravity. As long as the inertial components of heat transfer dominate the viscous and diffusive components, existing normal-gravity forced-convection heat-transfer correlations should be appropriate for calculating surface heat transfer. When the flow rates decrease sufficiently, to an order of magnitude comparable to the viscous and diffusive components, low-gravity effects should begin to dominate heat transfer. This heat-transfer regime could be important during startup, shutdown, or other operational transients expected in a space-based facility, but no experimental work is currently being performed in this area. In fact, these experiments may be difficult to perform on Earth because of the short duration of low gravity in terrestrial-based experimental facilities. Thus, these experiments may have to wait for the availability of a long-term experimental low-gravity laboratory such as the Space Shuttle or Space Station, or until long-term low-gravity terrestrial simulation techniques, such as magnetic levitation to offset gravity, are operational.

Boiling

Boiling and condensation are important heat-transfer regimes required for many power generation cycles. They also represent the most efficient method of transporting heat in any heat-removal system. Because boiling and condensation are two-phase phenomena, both are affected by the loss of a gravitational field. For example, boiling on Earth relies on the buoyant rise of vapor bubbles from a heated surface to supply the surface with a new source of liquid and prevent heater melting.

The growth of single bubbles in microgravity (10^{-4} to $4 \times 10^{-2} g$) has been studied by Cooper et al.[42] Water, toluene, and hexane have been separately examined under no-flow conditions. With the liquid in a saturated state, a single bubble was initiated at the wall by electrical means and its growth recorded with a high-speed camera. A simple relation was developed that governed the growth of diffusion-controlled bubbles. No sudden departure of the bubbles from the wall was observed; the lack of large temperature gradients was presumably responsible. The shapes of bubbles were found to be functions of surface tension, rate of growth, time, and the microgravity field. A relationship was also found between the maximum bubble diameter at departure from the wall and the gravitational field. Surface tension was observed to aid bubble departure by rounding off bubbles.

Another experimental study[43] examined the proportion of vapor bubble generation at the surface of saturated Refrigerant-11 (R-11) and bubble formation in the bulk liquid, upon venting to vacuum. The 5-s drop test facility was used with a photographic record plus some instrumentation such as pressure transducers and a thermistor. No bulk vapor was generated at $0 g$. All of the vapor was generated at the surface, but small amounts of bulk vapor, caused by boiling, were generated at measurable gravity levels. The vent rate, the percentage of vapor by volume, and the Bond number strongly influenced the amount of bulk vapor generated.

An earlier experimental program[44] studied forced-convection boiling at low heat flux and low velocities in microgravity. The liquid used was slightly subcooled ($0.4 - 1.5°C$) distilled water, heated from below with a flat chromel strip. Temperature was measured by a thermistor, and a 900 frame/s camera recorded the dynamics during the 2.2-s free fall. Bubble growth exhibited a cyclical trend; it is not clear if steady-state conditions prevailed. The majority (85%) of the bubbles remained attached to the heater surface, essentially forming a bubble boundary layer. The bubble diameter was found to correlate well with saturation layer thickness. The relevance of this work is probably restricted to storage tanks containing cryogenic fluids.

A very similar series of experiments by Cochran et al.[45] considered several fluids, but with no forced convection. The amount of subcooling was varied, as was the heat-transfer rate; the effects on bubble size and lifetime with gravity field were noted. The basic physical principles governing bubble dynamics were used in obtaining simple expressions for the dominant forces acting on the bubbles. These forces were calculated and plotted vs time.

Bubble size and lifetime in water were found to be nearly independent of the gravity field at high subcooling. For low subcooling, larger bubbles developed in 0 g than in 1 g. An ethanol/water solution, with a surface tension about 30% that of water, showed little influence of either g field or subcooling. The results are attributed to the more nearly spherical shape of the solution bubbles as compared to water. The pressure force therefore dominated solution bubble dynamics. A variation in heat-transfer rate (from 24,800 to 114,000 Btu/h-ft^2) for this solution also exhibited no trends with g field on bubble radii and lifetimes. Tests with a sugar/water solution having a viscosity 10 times that of water gave results similar to water regarding gravity and subcooling effects on bubble radii and lifetimes. The force histories for water and the sucrose solution are vastly different, however, with a significant drag force in the latter.

Oker and Merte[46] performed an elaborate series of pool boiling tests using liquid N_2 and Freon-113 and heat flux from 1×10^3 to 1×10^5 W/m^2. A short drop tower was used, which gave less than 1.4 s of free fall; it is uncertain whether steady-state conditions ever prevailed during the test. The g level was fairly high, up to 4×10^{-3} g. The data indicate that surface superheat at boiling inception is a function of gravity, and it is claimed to be less at 0 g than in normal gravity. However, an examination of the data shows that generally the temperature difference increased significantly in the transition from normal to 0 g, which is what one would expect, because buoyancy force driving natural convection heat transfer vanishes at 0 g. This study also shows a summary table listing earlier nucleate pool boiling reduced-gravity experiments and heat-transfer trends. These trends appear to be somewhat contradictory.

Experiments with pool boiling from wires and flat plates have been the subject of investigation since the early 1960's.[47] Dr. H. Merte at the University of Michigan is currently one of the prominent investigators of this phenomenon. Reference 47 summarizes the trends that relate to high- and reduced-gravity nucleate pool boiling.

Past experiments addressing commercialization activities also supply some information on bubble behavior in low gravity. A 4-min rocket flight provided the microgravity environment for an experiment on bubble migration in molten glass ("fining"), as reported by Wilcox et al.[8] A platinum heating strip melted a sample of sodium borate glass that contained entrapped voids. During the 0-g portion of the flight, distinct migration of the bubbles toward the hotter portion of the sample was noted. This observation is in agreement with the Brown model[7] of thermocapillary bubble migration, which predicts bubble motion against a thermal gradient.

Condensation

Similar to boiling, condensation is an important process required for space-based heat removal. Like boiling, condensation can be affected by the absence of gravity. Condensation in a terrestrial horizontal tube relies on the ability to limit the accumulation of condensate on the cooling surface by the gravitationally induced drainage of the condensate film. In low gravity, a condensing fluid like water with a relatively large liquid-surface attraction tends to accumulate on the condensing surface and form annular flow. In contrast, alkali metal fluids will be capable of condensing and removing heat because of their large heat-conduction capabilities. Nonmetallic fluids must rely on turbulence to transport heat through the film.

Experimental efforts with Freon-12[32-34] qualitatively indicated that condensation was unaffected by changes in gravity. Reduced-gravity experiments involving water condensation have been performed by F. Best of Texas A&M University. These experiments are discussed more fully in the section on Recent Experimental Efforts.

Critical Heat Flux

Critical heat flux concerns concentrate in two areas: pool boiling and forced-convection boiling. Each topic will be addressed separately.

Pool Boiling. Pool boiling in reduced gravity has been fairly well studied. Data for nucleate boiling from wires and flat plates[47] in the absence of gravity illustrate that temperature distribution will be altered sufficiently to affect both the nucleation and vaporization processes associated with boiling. A review article by Siegel[37] discusses and summarizes data on pre-1967 reduced-gravity experiments. Most experiments used the drop tower facilities, although some also utilized aircraft. For pool boiling, the critical heat flux between 0.01 and 1 g was found to correlate well with $g^{1/4}$. Whether this flux goes to 0 at 0 g could not be determined from the level and duration of the g field then achievable.

Forced-Convection Boiling. No reduced-gravity experimental data currently exist to relate critical heat flux to flow boiling. Past experimenters suggested that the critical heat flux value will decrease in low gravity, but experimental verification is lacking. Therefore, experimental work in this area is required.

Magnetic Field Tests

Several experiments have been performed with magnetic fields and a liquid metal (mercury). The objective was not to counteract gravity, but rather to note any perturbations engendered by the field on the boiling process. Faber and Hsu[48] applied a vertical magnetic field of 1–6 T to mercury undergoing nucleate pool boiling on a horizontal surface. Test results suggest that the magnetic induction encourages the incipience of boiling; that is, boiling can be initiated at a lower heat flux in its presence. It was postulated that the retarding influence of the Lorentz force increases bubble population and inhibits bubble motion (i.e., buoyancy is reduced) and agitation. Analysis indicated that the growing bubbles become elongated spheroids, with the major axis aligned with the magnetic field. These mechanisms were thought to be the chief contributors to the observed effects. In an earlier experiment (Hsu and Graham[49]), the magnetic field was oriented horizontally. Heat transfer was little perturbed in that configuration.

Petukhov and Zhilin[50] discuss a number of experiments performed with single-phase liquid metals in magnetic fields. Both transverse and longitudinal magnetic fields served to inhibit heat transfer. The effect was Reynolds number dependent, with the Nusselt number decreased as much as 30% at intermediate values of the Reynolds number. It was suggested that the magnetic fields affect the turbulence, but its exact structure (e.g., vortices in transverse fields) was not elucidated. In any case, these mechanisms may be relatively unimportant in two-phase flows.

The applicability of the experiments noted to the situation of interest (reduced gravity accomplished by means of a magnetic field) is probably remote. First, the action of various mechanisms was postulated, not proven. Second, an electric current passing through an alkali metal, causing it to flow vertically within a horizontal magnetic field, represents a significantly different situation from the pool boiling studies. Thus, the postulated mechanisms, even if valid, may be inoperable in a flow condition. All of these issues need to be investigated more fully; analytical studies and small experiments should prove or disprove the merit of magnetic fields as a means for generating reduced gravity.

Recent Experimental Efforts

Proposed Experiments — United States
A variety of fluids and fluids-related experiments are proposed in NASA documents.[51-54] Only a few two-phase flow experiments are actually in progress. Descriptions of these experiments follow. One difficulty with assessing current low-gravity experiments is lack of information; there is no central repository or clearinghouse on active programs.

The Shuttle program envisions a series of fluid mechanics experiments.[55] To this end, a drop dynamics module and a geophysical fluid flow cell have been constructed for use in the payload bay. The former module was flown in 1985, and tests studying the dynamics of rotating and oscillating free drops have been performed. No fluid flow experiments have been performed on the Shuttle to date.

Regarding future fluids experiments, many activities for the Shuttle and the space station deal with cryogen storage.[56] Very little from those studies will be applicable to two-phase turbulent fluid flow and heat transfer. One fluid experiment, however, will have a decided bearing on future two-phase activities. In 1985 the NASA Johnson Space Center (JSC) awarded a contract to engineer a Shuttle middeck experiment.[57] The objective is to obtain basic two-phase flow data; initial experiments will be adiabatic, using an air/water mixture. This experiment may be performed in cooperation with a university. It is anticipated that the experiment, which will take about 1.5 h, will be delayed several years past the date originally planned. An earlier, very detailed design proposed for the same experiment had features that were general enough to be suitable for incorporation into many future experiments.[58]

Recent and Continuing Experiments—United States

Texas A&M University. In the last few years the Nuclear Engineering Department of Texas A&M University has developed significant involvement in low-gravity fluids research. This involvement includes work for NASA-JSC and with other organizations.

This university has constructed a two-phase experiment for use on the JSC KC-135 aircraft.[59] The experimental module consists of two once-through boilers (one of which was supplied by Battelle, Pacific Northwest Laboratories) and a transparent condenser; heat (up to 5 kW) is supplied by nichrome heater wire wrapped around the boiler tube. Although there is some independent pressure control, the system operates essentially at atmospheric pressure as the condenser outlet vents overboard. There is also provision for adiabatic flow testing.

Preliminary results from the KC-135 flights indicate significant differences between observed condenser low-gravity flow regimes and 1-*g* data, when plotted on Quandt's flow regime map.[59] These tests used boiling and condensing water, or water/nitrogen in the adiabatic mode. Additional tests, with other fluids and over a broader parameter range, were recommended.

These low-gravity data were also plotted on 1-*g* flow regime maps created by various authors; these maps exhibit serious differences among themselves. The data agreed best with Dukler's 1-*g* map, although even on this map, only a little more than half of the annular flow data fall in the annular flow region.[60] Starting from theoretical force balances, the Texas A&M researchers also developed their own flow regime map consisting of three distinct flow patterns: dispersed, slug, and annular flow.[61] This map appears to be in general agreement with the data, but again more experiments, which appear to be forthcoming, are required to verify the flow regime transition lines completely. The university is slated to aid in the development of a test rig and subsequently construct and flight test it in a comprehensive program of flow regime mapping and pressure drop measurement.[62,63] One constraint on this effort is that all experiments will be adiabatic.

The completed experiments, performed on the NASA KC-135 aircraft, illustrate the dominance of the annular and bubbly flow regimes for condensation over a range of flow rates and void fractions. Figure 4 compares flow patterns in 0-*g* and horizontal 1-*g* tube condensers. Predictive calculations using the COBRA/TRAC computer code[64,65] with a normal-gravity vertical flow regime map and interfacial heat- and mass-transfer correlations predict larger condensation capabilities than experimental results for all compared cases except for the lowest total flow rate value (~ 100 kg/m²-s). Thus, it can generally be concluded that condensing heat transfer is better in a gravitational field unless a specialized design is introduced to enhance turbulence within the annular liquid film in low gravity.

A comprehensive discussion on condensation in low gravity, which expands on the Texas A&M work, is included in Chapter 7.

Battelle, Pacific Northwest Laboratories. Battelle, Pacific Northwest Laboratories (PNL), at the invitation of Texas A&M University, participated

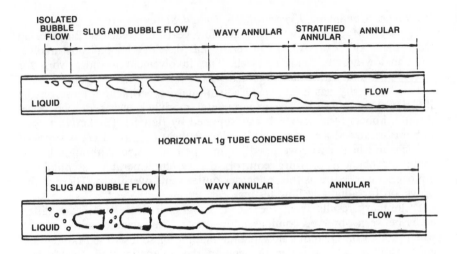

Fig. 4   Contrasting of flow regimes in tube condensation.

in four of the KC-135 flights. PNL built a transparent quartz boiler with appropriate instrumentation so that boiling behavior could be observed along with the condensation phenomena. The main motivator for studying low-gravity boiling was the need for analytical tools (i.e., computer codes capable of detailed thermal-hydraulic modeling) to design nuclear reactor space power systems and assess their performance.

When the boiler and condenser test data are plotted on the revised flow regime map developed by Dukler,[66] they fit quite well within the regime transition boundaries.[65] The three types of flow regimes observed, bubbly, slug, and annular, replicate those of Dukler with only a single questionable data point.

These experiments[67,68] have shown that in reduced gravity an annular flow regime appears to be dominant for forced water flow in circular tubes. This flow regime is beneficial in a boiler because the surface tension attraction between a wetting fluid and its container would help to keep the heating surface wet and promote heat transfer. The experimental results for a large range of flow rates and heat fluxes also indicate that sufficient inertial forces exist in the annular liquid layer to promote rapid movement of the vapor bubbles from the heater surface. This results in a situation similar to flow patterns expected in a vertical evaporator tube (Fig. 5). In 1 *g*, with liquid inlet velocities greater than about 1 m/s, the flow patterns are not strongly dependent on tube orientation. However, for some conditions, the annular film observed in low gravity was thicker than that in 1 *g* and possessed surface waves or instabilities that sometimes bridged the annular gap (thick film annular-bubble slug flow). The temperature rise in the boiler for the reduced-gravity experiments was comparative but slightly lower than that obtained from the COBRA/TRAC computer code,[64,65]

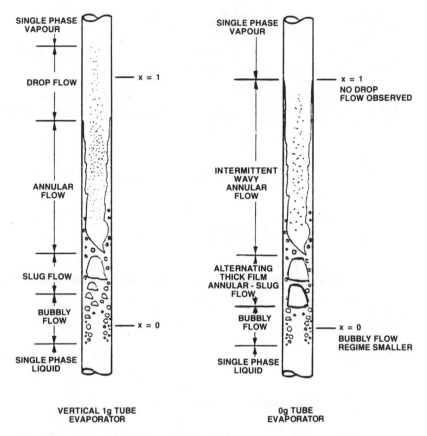

Fig. 5   Flow patterns in an evaporator tube ($V_{inlet} > 1$ m/s).

which used vertical flow rate correlations for boiling. It is expected that boiling flows at very small flow rates would tend to approach a pool boiling condition. Figure 6 compares flow patterns for a horizontal evaporator tube in 1 g and a 0-g evaporator tube both with inlet velocities less than about 1 m/s.

Both boiler and condenser performance were modeled with COBRA/TRAC, a computer code that provides a two-fluid, three-field representation of two-phase flow. Low gravity was simulated in the code by setting the gravity terms in the momentum equations to zero. Vertical boiling flow rate correlations are incorporated in the code. As anticipated, this simplistic approach hardly suffices for accurate results. Calculated values of boiler conditions (temperature and pressure) are generally in good to fair agreement with the measured data; predictions of condenser performance are usually far off, especially the pressure drop, which was grossly underpredicted. These discrepancies occurred even when the code determined the

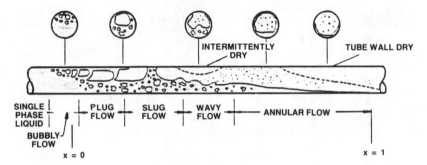

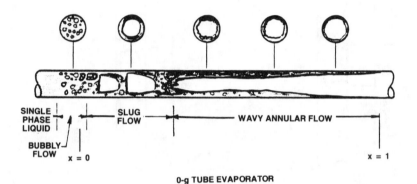

Fig. 6　Flow patterns in an evaporator tuve ($V_{\text{inlet}} > 1$ m/s).

flow pattern correctly. Clearly, additional low-gravity data and correlations are required for realistic modeling of two-phase flows.

Flow boiling phenomena in low gravity are discussed more fully in Chapter 6.

University of Kentucky. The University of Kentucky, under NASA-JSC sponsorship, has performed work in the area of microgravity condensation.[69-71] To date, the published work has only addressed analytical efforts, including a literature review of relevant experiments.[69] Improved heat transfer by means of condensate removal using vapor shear mechanisms has received considerable attention. In addition, the use of high-strength, nonuniform electric fields (dielectric forces) for condensate removal has also been studied.

AETA Corporation. An interesting approach toward simulating low gravity has been pursued by the AETA Corporation through the use of two immiscible fluids of similar density, water and polypropylene glycol, to approximate low-gravity two-phase flow.[72] However, serious doubts exist regarding the validity of this approach[66,73] because it is obvious that only a partial similitude to space conditions is attainable by this technique.[5]

University of Houston. Under the aegis of the NASA Lewis Research Center (LeRC), the University of Houston has been examining adiabatic flow regimes in low-gravity drop tower and Learjet experiments.[66] Thus far, three flow regimes appear to characterize all low-gravity flows: bubbly, slug, and annular. It has been conjectured that the bubbly and slug flow regions are not truly separate, but represent points on a continuum rather than a transition from one physical process to another. A much more detailed discussion in adiabatic flow regimes merits an entire chapter and is included in Chapter 5.

Sundstrand Corporation. With funding from NASA-JSC, the Sundstrand Corporation has built and tested a two-phase heat-transfer loop incorporating its rotary fluid management device (RFMD) in low-gravity parabolic trajectories in the NASA KC-135 aircraft.[74,75] This RFMD separates the liquid and vapor flow components and is specifically designed for space operation.[76] The test section consisted of two transparent straight tubes, 1.83 m long × 15.8 mm i.d., connected by two 45-deg elbows and a flexible line. The total length of the test section was 6.7 m and included additional fittings, valves, and bends. The quality of the working fluid (R-114) was varied by means of electrical heaters of approximately 5-kW capacity. The test section inlet quality was varied from about 0.05 to 0.90; flow regimes were filmed near the exit of the second transparent section.

Only two flow regimes were observed during testing: slug and annular flow. Test data were plotted on various flow regime maps; all were found to be deficient in predicting the observed flows, although often this could be remedied with a minor adjustment of map parameters.[76]

Low-gravity pressure drop measurements could be made to correlate well with calculated values when use was made only of observed, rather than predicted, flow regimes. One incontrovertible fact that became very evident is that low-gravity pressure drop is consistently higher (by about 30–50%) than that in 1 g.[76] The shifted flow patterns in low gravity were presumably responsible for this increased pressure loss.

University of Michigan. The University of Michigan continues its involvement in low-gravity pool boiling research and is extending its efforts to the study of flow boiling.

National Bureau of Standards. The National Bureau of Standards (NBS) has been developing a holographic interferometry technique for measuring the temperature profile in a thermal boundary layer.[77] This technique, which uses a diffuse light source and double exposures, was used in both 1-g and KC-135 low-gravity experiments to measure the steep temperature gradients in transient heat transfer. The experimental apparatus consisted of a small titanium foil heater plate, a platinum thermometer for determining the bulk temperature of the R-13 heat-transfer fluid, and the interferometric equipment.

The normal-gravity results indicate that approximately three-fourths of the boundary layer can be probed holographically. However, an unexplained discrepancy exists between theoretical results and observations at fringe locations. The low-gravity experiments were somewhat limited because of the low-powered laser used on the KC-135. The loss of image resolution resulted in only occasional observation of the fringe patterns. Further low-gravity tests should aid in optimizing this optical technique.

Ford Aerospace/McDonnell Douglas Corporation. A combined analytical and experimental study has been conducted to improve modeling of liquid-propellant behavior in low gravity.[78] Initial tests utilized a drop tower that permitted 2 s of low gravity. The test vessel was a 3.6-in.-diam sphere. The behavior of colored distilled water was recorded with a movie camera. A similar KC-135 experiment consisted of a two-tank system that was free floated in the cabin. Measured parameters included liquid pumping capability, the time to reach an equilibrium state, and the stable equilibrium liquid free-surface shape.

A computer program was developed to study the influence of tank geometry on the liquid free surface, the volume fill fraction, the Bond number, and the contact angle. Agreement between test and analysis was found to be good. Other programs analyzed transient liquid reorientation characteristics and capillary rise time and pumping capability in channels of various cross sections. Excellent agreement was found between tests and analyses.

Propellant settling and slosh analyses were performed using the SOLA-VOF computer code. This code appeared capable of analyzing tank fluid dynamics problems but was not calibrated against test data in this study.

Of related interest was an analytical study that examined the equilibrium fluid interfaces in the absence of gravity.[79] The mathematical results indicated that these interfaces are dependent on a critical angle, which is the sum of the contact angle between the fluid interface and the container wall, and the interior half angle if the container consists of a regular polygon. For the case of a cylinder of general cross section, a somewhat more complex formulation is required that again shows the existence of a unique critical angle for a stable fluid equilibrium shape.

Recent and Continuing Experiments — Europe

Paradoxically, although motivated by the commercial aspects of improved materials processing in space, European low-gravity experiments have a distinctly "basic physics" flavor. No doubt this is partly inherent in the complex phenomena encountered in the processing of exotic materials, but in part it must be ascribed to some philosophical differences between the U.S. and European approaches to science and its applications.

Another difference in methodology is evident from European use of a broader range of low-gravity environments. These range from Spacelab,[80,81] through parabolic flights and sounding rockets,[82,83] to experiments released from balloons.[84] Overall, the Europeans exhibit a very serious attitude toward obtaining the desired data in a methodical cost effective manner.[85] The development of the fluid physics module (FPM), a multipurpose apparatus for research on low-gravity fluid behavior, reflects this seriousness.[86] The European low-gravity experiments receive only brief coverage here, not because they lack merit, but because at this time the topics they have explored are of only peripheral interest to the design of large space-based thermal-hydraulic systems.

Marangoni flows have received considerable attention,[80,87] including investigation of oscillatory/turbulent transition during a TEXUS rocket flight[88] and laminar/oscillatory transition in a Spacelab experiment.[89] A more recent experiment flown on the D1 Spacelab mission continued the investigation begun in the first mission and included subsidiary KC-135 flight experiments. Results were in good agreement with earlier data,[90,91] and it was concluded that oscillatory behavior in Marangoni flow is strongly geometry dependent and may be influenced by any residual gravity present.[90] Yet another conclusion reached on the D1 mission was that a description of motion at a liquid/gas interface is incomplete if it is solely in terms of the surface tension and its temperature dependence.[92] An additional parameter, the dilational surface viscosity, was recommended. Perhaps of equal interest here are the experiments designed to examine other interfacial phenomena, such as wetting,[93] capillarity and diffusion (performed in the FPM),[94] and similar phenomena.

The study of pool boiling in Freons has been another topic of considerable interest.[72,83] With saturated boiling taking place on a flat plate in sounding rocket experiments, it was found that forces holding a bubble to a surface dominated over those promoting detachment. This conclusion is not valid for arrangements with small bubble-heater attachment surfaces such as exist with thin wires. It was also concluded that microgravity steady-state nucleate pool boiling is impossible, and film boiling becomes established very rapidly after boiling onset.[72] Steady-state nucleate boiling of a subcooled liquid was observed. Heat transfer was almost unaffected by the presence of microgravity conditions up to about one-half the terrestrial critical heat flux value. However, the bubble population and the number of active sites decreased because the bubble radii increased by a factor of about 20.[72]

Experimental Summary

A summary of the available low-gravity fluid-thermal experiments is shown in Table 10. This short list represents the currently available knowledge of low-gravity thermal-hydraulics and exemplifies the need for additional experimentation. Using these data, preliminary trends for fluid-thermal behavior can be predicted (Table 11). Few quantitative relations are available or verified. The only exceptions may be in the areas of flow regime determination and pool boiling where experimental efforts are proceeding and preliminary relations are known.

Reduced-Gravity Experimental Facilities

The previous section discussed the experimental data needed to develop reduced-gravity fluid-thermal relations. Experiments to supply the required data can be performed in Earth-based or space-based facilities.[95] Space-based research facilities are very limited. As expected, the Earth-based experimental facilities are cheaper to operate than space-based facilities (i.e., the Space Shuttle). However, they possess operational limitations

Table 12 Experimental facilities and techniques for attaining a reduced-gravity environment

Type	Features	Special conditions	Lowest g level attainable[a]	Reduced-gravity duration, s	Comment
Drop	Various towers/tubes (13.2–145 m high)	Vacuum/drag shield cryogenic	$\leqslant 1 \times 10^{-5}$	2–5 (10, if accel. from bottom	Time is severe constraint; high deceleration rate ($\sim 30\,g$) to stop
Magnetic,[b] viscous, sonic, inertial	"Levitate" sample	—	Unknown	Days	Techniques may generate secondary effects, disturbing or distorting sample
Aircraft	Repeated parabolic trajectories between two altitudes	Can accommodate large payloads	$\leqslant 0.01$	~ 20	Relatively high g field level; difficult to maintain at steady value
Rocket	Free-fall mode	—	Unknown	~ 250	No details provided in literature[c]
Spacelab (Shuttle)	Limited size of experimental package	Stringent safety requirements	1×10^{-2} to 1×10^{-6} (jitter free on c.g.)	Days	Long scheduling lead time; crew members can be used to run experiments

[a]Fraction of the Earth's gravity.
[b]Although these techniques generally do not provide truly 0-g conditions, they may be of some utility here and are included for completeness. Gravity fields > 1 can be attained by inertial techniques, whereas near-0-g conditions are possible through use of magnetic fields.
[c]The European TEXUS program is well documented; see the section on Recent and Continuing Experiments—Europe.

(time duration and steadiness) that must be accounted for in the experiment performance or in the subsequent data reduction and interpretation. Small, low-power experiments have been performed in the Shuttle using nonhazardous fluids. The primary available Earth- and space-based experimental facilities are listed and discussed here and are summarized in Table 12.

Earth-Based Reduced-Gravity Experimental Facilities

Drop Towers and Tubes

Drop tower and tubes are structures built for studying reduced-gravity effects. They consist of a tower or shaft in which an experimental package is dropped from an elevated position and subsequently arrested at a lower one. During the intervening "free-fall" period in the experimental phase, a reduced-gravity environment is present within the experimental module. In this period, an object freely falling in a gravitational field has no net forces acting within it and therefore replicates the phenomena one would observe if that object had no external forces whatsoever acting on it.

Several deficiencies are associated with these facilities. Because the free-fall experiment is constantly accelerating at 9.8 m/s^2, there is an evident limit to a reasonably sized tower or tube. This severely constrains the duration of the experiment. Aerodynamic drag imposes a net force on the experiment. Low-gravity levels (about $10^{-6}g$) can be achieved by evacuating the air from the drop tube before an experiment or by employing drag shields in nonevacuated towers.

Although various organizations in addition to NASA possess drop towers, these towers are generally small with short reduced-gravity periods. The most accessible facilities are at the NASA-LeRC. At LeRC, a 2.2-s duration nonevacuated drop tower (Fig. 7) and a 5.2-s duration evacuated drop tube are available for experiments.[96,97] The nonevacuated tower offers quick turnaround so that as many as nine experiments can be performed each day. Its drawbacks are the short free-fall time and the relatively higher-gravity environment resulting from aerodynamic drag. The 5.2-s tube provides additional time, but because the entire tube must be pumped out (to $\sim 10^{-2}$ torr) between tests, only two or three experiments per day are feasible. Furthermore, each facility has unique requirements that preclude construction of a single test vehicle acceptable to both the drop tower and tube.

Another low-gravity drop experimental method using a balloon has been tested recently. The balloon drops an experimental package from a high elevation, about 25 miles, to obtain a free-fall condition in a manner similar to a drop tower or tube. Problems exist in designing an adequate drag shield and an appropriate device to arrest the experiment package from the free-fall condition. This procedure has recently been tested at the Kiruna test range in Sweden.[98] The Mikroba 2 test capsule has been reported to provide up to 60 s of microgravity.

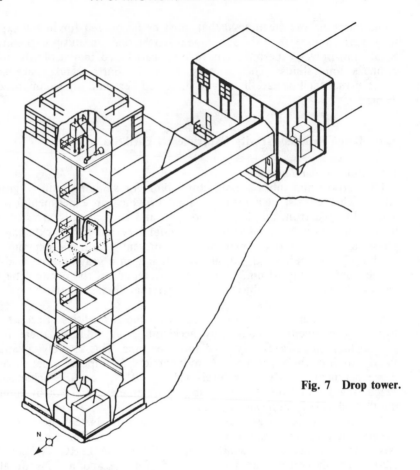

Fig. 7 Drop tower.

Low-Gravity Research Aircraft

An aircraft flying a parabolic trajectory can attain low-gravity levels. These levels can be sustained considerably longer than those in drop towers because the elevation change during free fall is much greater. The aircraft also offers significantly more room for an experiment, and power can be drawn from onboard generators, obviating the need for batteries. An additional advantage is that the experimenter can fly along with the experiment and is able to both observe and control the test while it is in progress.

Disadvantages are also associated with this mode of reaching reduced gravity. The magnitude and uniformity of the g level produced is dependent on many factors, the chief one being the skill of the pilot. Before the free-fall maneuvers, an acceleration of about $2\,g$ is experienced. This acceleration may perturb flow in the test loop, increasing the time required to establish steady state once reduced gravity is achieved. Thus, the period for taking valid data may be shorter than the free-fall time.

NASA owns two aircraft dedicated to reduced-gravity research. A Learjet available at LeRC is capable of operating for 15–20 s in a free-fall mode[84] and can fly up to six free-fall trajectories per mission. A much larger aircraft, the KC-135, is similarly maintained by the JSC (Fig. 8).[99,100] Somewhat longer free-fall conditions, about 25 s, can be obtained and as many as 40 low-gravity maneuvers can be performed on a single flight. This aircraft also affords the opportunity to perform experiments either bolted down or free floated within the cabin. The free-float procedure provides a better low-gravity condition since it eliminates extraneous acceleration imparted by the aircraft.

Rockets

A rocket can perform essentially the same function as the aircraft just described. A Space Processing Application Rocket (SPAR VIII) used in a fluids experiment[8] provided more than 4 min of an "average acceleration near zero." The actual *g* level was not reported. Because of the hard landing, it appears that only the film record survived intact.

Additional difficulties are encountered in the use of a rocket for reduced-gravity experiments. Survival and recovery of the experimental module is certainly an issue, and the size of this module is quite restricted by the rocket itself. The transition from high to reduced gravity, when compared to aircraft parabolas, is accentuated here. It is not clear how low and steady a *g* level is within rocket capabilities. Finally, the number of rockets that

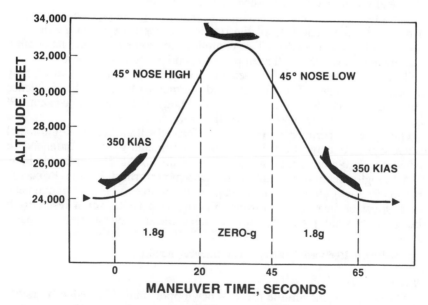

Fig. 8 KC-135 aircraft trajectory.

would have to be expended in any realistic test series could be too large to be practical.

Magnetic Levitation

Without a doubt, garnering data on two-phase behavior in low gravity by means of the facilities described represents an arduous and complex approach. If possible, one would like to work directly with the fluid of interest on Earth and still obtain valid data. One potential means for doing so is to investigate behavior in high-gravity (>1 g) fields created mechanically by a centrifuge, for example. Given data points at 1 g, and various higher gravity levels, one is tempted to extrapolate to a 0-g condition. This extrapolation is never a reliable technique and can lead to serious error. Extrapolation in this manner to the singularity of a 0-g situation is not justifiable; there is simply no way to evaluate the results so achieved.

One alternative exists. The fact that liquid alkali metals are excellent conductors of electricity can be used to advantage. A magnetic field can be employed to "levitate" the alkali metal. By proper orientation, the magnetic field can cancel the gravitational body force, producing essentially 0-g conditions. Of course, the uniformity of this field must be ensured and Joule heating of the alkali metal must be minimized.

A well-known relationship exists between a magnetic field, an electric current, and a moving conductor. The geometric relationship among these three parameters is illustrated by the familiar right-hand rule.[101] An electric motor is a typical example of the motion induced in the conductor (armature) by an electric current in a magnetic field.[102] A less familiar but more apropos example is the electromagnetic pump specifically developed for pumping electrically conductive fluids such as alkali metals.[103]

For a test section, one can envision a vertical length of pipe entirely within a horizontal magnetic field of the requisite strength to cancel gravity and provide adequate mass flow. Two complications come to mind: 1) assuming heating in the test section, the temperature, and properties, of the alkali metal will vary over the length of the section; and 2) the generation of vapor may introduce anomalies. These effects, and the manner in which they influence the desired cancelation of the gravitational body force, need to be examined in detail to assess the feasibility of this approach.[1]

Only a few experiments have been performed with a boiling liquid metal (mercury) in a magnetic field.[44] These experiments were not intended to investigate a 0-g condition; hence, their results are hardly applicable here.

Although preliminary studies have been performed by the Princeton Plasma Physics Laboratory in developing this approach,[1] the use of magnetic fields must be investigated further. Enormous benefits would result from Earth-based alkali metal experiments that replicate the conditions in space.

Space-Based Reduced-Gravity Experimental Facilities

Space Shuttle

The advent of the Shuttle presents new opportunities for reduced-gravity and other space-related experiments.[104] Instruments and small experiments can be housed in small storage containers or in the crew's storage lockers. Larger, more complex test apparatus can be rack-mounted at the aft flight deck or hard mounted to the Shuttle structure within the payload bay. With the remote articulator system (RMS), a payload can be maneuvered outward to distances of 15 m.

These features are merely embellishments to the Shuttle's unique capability to maintain a reduced-gravity environment (about $10^{-4}g$) for long periods. Thus, Shuttle-based fluids experiments can be run until steady-state conditions prevail, and the low unvarying g-level ensures the absence of gravity-dominated phenomena. Yet another advantage of the Shuttle over most other facilities is the possibility of involving crew members in operating or controlling the tests.

A full description of Shuttle capabilities and requirements must encompass the less attractive features as well. The requirements for an experimental package are rather onerous in terms of development costs, safety considerations, and, of course, the expense of the Shuttle flight itself. Any and every experimenter is faced with clearing these time-consuming hurdles. The extremely conservative safety criteria, meant to guarantee the safety of the crew, preclude the presence of any hazardous materials or conditions in an onboard experiment.

Although it represents a significant improvement over the other facilities, the Shuttle environment is by no means ideal. The g level typically present during orbit is 10^{-3}–10^{-4}; jitter can degrade it to 10^{-2}. A true microgravity condition ($10^{-6}g$) is achieved only at the Shuttle's center of gravity.[105] Accelerometer measurements taken during the D1 Spacelab (Shuttle) mission graphically illustrate the typical perturbations resulting from various activities.[106] In comparison, the space station of the 1990's has a 10^{-5}- to 10^{-6}-g level requirement. The space station, which is still being designed, will represent the first true in-space laboratory with extensive power, instrumentation, data acquisition, and computational capabilities.[107] But until its debut, experimenters must look to the Shuttle for the best reduced-gravity environment.

Summary and Conclusions

The design of advanced fluid-thermal systems for space application will require the development of constitutive thermal-hydraulic models and analysis techniques that account for the microgravity environment. At present, only approximate methods of performing two-phase low-gravity thermal-hydraulic calculations exist. A more extensive experimental data base is required to serve as the basis for understanding low-gravity fluid-thermal phenomena and for developing appropriate low-gravity analytical models. Thus, Earth- and space-based thermal-hydraulic experiments and concurrent constitutive model and analytical technique development efforts are needed to ensure the design optimization and safety of all advanced space power, propulsion, and thermal management systems.

References

[1]Krotiuk, W. J. and Antoniak, Z. I., "Thermal-Hydraulic Assessment Report for the Multimegawatt Space Nuclear Power Program," Vols. 1 and 2, Battelle, Pacific Northwest Lab., Richland, WA, Rept. PNL-6061, Oct. 1986 (Draft).

[2]Krotiuk, W. J. and Antoniak, Z. I., "The Development of Two-Phase Computer Codes for Microgravity Applications," *Symposium on Microgravity Fluid Mechanics*, American Society of Mechanical Engineers, Fluids Engineering Division, Vol. 42, New York, Dec. 1986.

[3]Krotiuk, W. J., "Two-Phase Computer Codes for Zero-Gravity Applications," *Advances in the Astronautical Sciences*, Vol. 64, Pt. III, American Astronautical Society, March 1987.

[4]Antoniak, Z. I., "Two-Phase Alkali Metal Experiments in Reduced Gravity," Battelle, Pacific Northwest Lab., Richland, WA, Rept. PNL-5906, June 1986.

[5]Sheahen, T. P., "Physical Principles of Microgravity Research," NASA Office of Space Commercialization, Western Technology, Inc., Derwood, MD, Feb. 1986.

[6]Ostrach, S., "Low-Gravity Fluid Flows," *Annual Review of Fluid Mechanics*, Vol. 14, 1982, pp. 313–345.

[7]Collier, J. E., *Convective Boiling and Condensation*, McGraw-Hill, New York, 1972, pp. 18, 35, 165.

[8]Wilcox, W. R., Subramanian, R. S., Meyyappan, M., Smith, H. D., Mattox, D. M., and Partlaw, D. P., "A Preliminary Analysis of the Data from Experiment 77-13 and Final Report on Glass Fining Experiments in Zero Gravity," NASA CR-161884, 1981.

[9]Cullimore, B. A., Goble, R. G., Jensen, C. L., and Ring, S. C., "SINDA '85/FLUINT Systems Improved Numerical Differencing Analysis and Fluid Integration," Version 2.1, Martin Marietta, Rept. MCR-86-594, Nov. 1987.

[10]Carlson, K. E., Roth, P. A., and Ransom, V. H., "ATHENA Code Manual," Idaho National Engineering Lab., EG&G Idaho, Idaho Falls, ID, Rept. EGG-RTH-7379, Sept. 1986.

[11]Carlson, K. E., Ransom, V. H., and Roth, P. A., "ATHENA Solutions to Development Assessment Problems," Idaho National Engineering Lab., EG&G Idaho, Idaho Falls, ID, Rept. EGG-RTH-7651, March 1987.

[12]Shumway, R. W., "ATHENA Radiation Model," Idaho National Engineering Lab., EG&G Idaho, Idaho Falls, ID, Rept. EGG-TFM-7925, Oct. 1987.

[13]Thurgood, M. J. and Kelly, J. M., "COBRA/TRAC—A Thermal-Hydraulic Code for Transient Analysis of Nuclear Reactor Vessels and Primary Coolant Systems," Vols. 1–5, Battelle, Pacific Northwest Lab., Richland, WA, Rept. NUREG/CR-3046, March 1983.

[14]Navickas, J. and Rivard, W. C., "Applicability of the FLOW-NET Program to Solution of Space Station Fluid Dynamics Problems," *International Symposium on Thermal Problems in Space-Based Systems*, Vol. 83, American Society of Mechanical Engineers, New York, Dec. 1987.

[15]Campbell, J. R., Rivard, W. C., and Sicilian, J. M., "FLOW-NET: A Program for Water-Steam-Air Thermal-Hydraulics in Pipe Networks," Flow Science, Los Alamos, NM, Rept. FSI-82-00-1, May 1982.

[16]Taylor, D. D., "TRAC-BD1: An Advanced Best Estimate Computer Program for Boiling Water Reactor Transient Analysis," Vols. I–III, Idaho National Engineering Lab., EG&G Idaho, Idaho Falls, ID, Rept. NUREG/CR-3633, 1984.

[17]Weaver, W. L., Shumway, R. W., Singer, G. L., and Rouhani, S. Z., "TRAC-BF1 Manual: Extensions to TRAC-BD1/MOD1," (Draft), Idaho National Engineering Lab., EG&G Idaho, Idaho Falls, ID, Rept. NUREG/CR-4391, 1985.

[18]Sicilian, J. M. and Hirt, C. W., "HYDR3D, A Solution Algorithm for Transient 3D Flows," Flow Science, Los Alamos, NM, Rept. FSI-84-00-3, 1985.

[19]*FLOW-3D, Computational Modeling Power for Scientists and Engineers*, Flow Science, Inc., Los Alamos, NM, Technical Manual.

[20]Torrey, M. D., Cloutman, L. D., Mjolsness, R. C., and Hirt, C. W., "NASA-VOF2D: A Computer Program for Incompressible Flows with Free Surfaces," Los Alamos National Engineering Lab., Los Alamos, NM, Rept. LA-10612-MS, Dec. 1985.

[21]Torrey, M. D., Mjolsness, R. C., and Slein, L. R., "NASA-VOF3D: A Three-Dimensional Computer Program for Incompressible Flows with Free Surfaces," Los Alamos National Engineering Lab., Los Alamos, NM, Rept. LA-11009-MS, July 1987.

[22]George, T. L., Basehore, K. L., Wheeler, C. L., Trather, W. A., and Masterson, R. E., "COBRA-WC: A Version of COBRA for Single-Phase Multiassembly Thermal-Hydraulic Transient Analysis," Battelle, Pacific Northwest Laboratories, Richland, WA, Rept. PNL-3259, 1980.

[23]Rector, D. R., Cuta, J. M., Lombardo, N. J., Michener, T. E., and Wheeler, C. L., "COBRA-SFS: A Thermal-Hydraulic Analysis Computer Program," Vols. I–III, Battelle, Pacific Northwest Lab., Richland, WA, Rept. PNL-6049, Nov. 1986.

[24]"COMMIX-1B: A Three Dimensional Transient Single-Phase Computer Program for Thermal-Hydraulic Analysis of Single- and Multicomponent Systems," Argonne National Lab., Argonne, IL, Rept. NUREG/CR-4348, 1985.

[25]"FLUENT, Computer Programs for Simulating Fluid Flow," Creare, Hanover, NH, Seminar Notes, Oct. 1987.

[26]Gunton, M. C., "PHOENICS, An Instructional Manual," Concentration Heat and Momentum Limited, Rept. CHAM TR/75, 1983.

[27]Trent, D. S., Eyler, L. L., and Budden, M. J., "TEMPEST, A Three-Dimensional Time-Dependent Computer Program for Hydrothermal Analysis," Vols. 1 and 2, Battelle, Pacific Northwest Lab., Richland, WA, PNL-4348, Sept. 1983.

[28]Wheeler, C. L., Thurgood, M. J., Guidotti, T. E., and De Bellis, D. E., "COBRA-NC: A Thermal-Hydraulic Code for Transient Analysis of Nuclear Reactor Components," Vols. 1, 2, 4, and 7, Battelle, Pacific Northwest Lab., Richland, WA, Rept. NUREG/CR-3263 (PNL-5515), 1986.

[29]"COMMIX-2: A Three-Dimensional Transient Computer Program for Thermal-Hydraulic Analysis of Two-Phase Flows," Argonne National Lab., Argonne, IL, Rept. NUREG/CR-4371 (ANL-85-47), 1985.

[30]"TRAC-PF1/MOD1—An Advanced Best-Estimate Computer Program for Pressurized Water Reactor Thermal-Hydraulic Analysis," Los Alamos National Lab., Los Alamos, NM, 1984.

[31]Heppner, D. B., King, C. D., and Littles, J. W., "Zero-G Experiments in Two Phase Fluids Flow Regimes," presented at the Intersociety Conference on Environmental Systems, San Francisco, CA, July 21–24, 1975.

[32]Williams, J. L., "Space Shuttle Orbiter Mechanical Refrigeration System," NASA CR-144395, Oct. 1974.

[33]Keshock, E. G., Spencer, G., French, B. L., and Williams, J. L., "A Photographic Study of Flow Condensation in 1-G and Zero-Gravity Environments," presented at the Fifth International Heat Transfer Conference, Tokyo, Japan, Sept. 3–7, 1974.

[34]Williams, J. L., Keshock, E. G., and Wiggins, C. L., "Development of a Direct Condensing Radiator for Use in a Spacecraft Vapor Compression Refrigeration System," *Journal of Engineering for Industry*, Vol. 95, Nov. 1973, pp. 1053–1064.

[35]Albers, J. A. and Macosko, R. P., "Experimental Pressure-Drop Investigation of Nonwetting, Condensing Flow of Mercury Vapor in a Constant-Diameter Tube in 1-G and Zero-Gravity Environments," NASA TN D-2838, June 1965.

[36]Namkoong, D., Block, H. B., Macosko, R. P., and Crads, C. C., "Photographic Study of Condensing Mercury Flow in 0- and 1-G Environments," NASA TN D 4023, June 1967.

[37]Siegel, R., "Effects of Reduced Gravity on Heat Transfer," *Advances in Heat Transfer*, Vol. 4, Academic, New York, 1967, pp. 143–228.

[38]Naumann, R. J., "Materials Processing in Space: Review of the Early Experiments," *Space Science and Applications*, IEEE Press, New York, 1986, pp. 159–171.

[39]Oran, W. A., "Current Program to Investigate Phenomena in a Microgravity Environment," *Space Science and Applications*, IEEE Press, New York, 1986, pp. 173–175.

[40]Carruthers, J. R., "Materials Science and Engineering in Space," *Space Science and Applications*, IEEE Press, New York, 1986, pp. 155–158.

[41]Rezkallah, K. S., "Two-Phase Flow and Heat Transfer at Reduced Gravity: A Literature Survey," presented at the ASME Heat Transfer Conference, Houston, TX, July 1988.

[42]Cooper, M. G., Judd, A. M., and Pike, R. A., "Shape and Departure of Single Bubbles Growing at a Wall," presented at the Sixth International Heat Transfer Conference, Toronto, Canada, Aug. 7–11, 1978.

[43]Labus, T. L., Aydelott, J. C., and Lacovic, R. F., "Low-Gravity Venting of Refrigerant 11," NASA TM X-2479, Feb. 1972.

[44]Cochran, T. H., "Forced-Convection Boiling Near Inception in Zero Gravity," NASA TN D-5612, Jan. 1970.

[45]Cochran, T. H., Aydelott, J. C., and Spuckler, C. M., "An Experimental Investigation of Boiling in Normal and Zero Gravity," NASA TM X-52264, 1967.

[46]Oker, E. and Merte, H., Jr., "Transient Boiling Heat Transfer in Saturated Liquid Nitrogen and R113 at Standard and Zero Gravity," NASA CR-120202, Oct. 1973.

[47]Merte, H. M., "Nucleate Pool Boiling: High Gravity to Reduced Gravity; Liquid Metals to Cryogenics," *Transactions of the Fifth Symposium on Space Nuclear Power Systems*, Albuquerque, NM, CONF-880122-Summs, Jan. 1988.

[48]Faber, O. C., Jr. and Hsu, Y. Y., "The Effect of a Vertical Magnetic Induction in the Nucleate Boiling of Mercury Over a Horizontal Surface," *Chemical Engineering Progress Symposium Series*, Vol. 64, American Institute of Chemical Engineers, New York, 1968, pp. 33–42.

[49]Hsu, Y. Y. and Graham, R. W., *Transport Processes in Boiling and Two Phase Systems*, Hemisphere, Washington, DC, 1976, Chaps. 7–10.

[50]Petukhov, B. S. and Zhilin, V. G., "Heat Transfer in Turbulent Flow of Liquid Metals in a Magnetic Field," *Heat Transfer in Liquid Metals, Progress in Heat and Mass Transfer*, Vol. 7, Pergamon, New York, 1973, pp. 553–568.

[51]NASA, "Microgravity Science and Applications," brochure prepared by the Marshall Space Flight Center, Huntsville, AL (no date).

[52]Pentecost, E. A., "Materials Processing in Space," NASA TM-82525, 1983.

[53]Pentecost, E. A., "Microgravity Science and Applications Bibliography—1984 Revision," NASA TM-86651, Sept. 1984.

[54]Naumann, R. J., "Microgravity Science and Application Program Description Document," NASA Marshall Space Flight Center (MSFC), AL, 1982.

[55]NASA, "Spacelab 3," EP203, 1984.

[56]NASA, "In-Space Research, Technology, and Engineering Workshop," Williamsburg, VA, Oct. 8–10, 1985.

[57]Schuster, J. R., "Liquid-Vapor Flow Regimes in Microgravity Experiments," presented at the NASA In-Space Research, Technology, and Engineering Workshop, Williamsburg, VA, Oct. 8–10, 1985.

[58]Bradshaw, R. D. and King, C. D., "Conceptual Design for Spacelab Two Phase Flow Experiments," General Dynamics Convair Division, NASA CR-135327, Dec. 1977.

[59]Kachnik, L., Lee, D., Best, F., and Faget, N., "A Microgravity Boiling and Convective Condensation Experiment," American Society of Mechanical Engineers, Paper 87-WA/HT-12, 1987.

[60]Lee, D., Best, F. R., and McGraw, N., "Microgravity Two Phase Flow Regime Modeling," *Third Proceedings of Nuclear Thermal-Hydraulics*, American Nuclear Society, La Grange Park, IL, Nov. 1987, pp. 94-100.

[61]Lee, D. and Best, F. R., "Microgravity Two Phase Fluid Flow Pattern Modeling," *Transactions of the Fifth Symposium on Space Nuclear Power Systems*, Albuquerque, NM, CONF-880122-Summs, Jan. 1988.

[62]Best, F. R., personal communication, Texas A&M University, College Station, TX, May 11, 1988.

[63]Hill, W., personal communication, Foster Miller Co., Waltham, MA, May 23, 1988.

[64]Cuta, J. M., Krotiuk, W. J., and Samuels, J. W., "Analysis and Interpretation of Low Gravity Boiling Experiments in the KC-135," *Transactions of the Fifth Symposium on Space Nuclear Power Systems*, CONF-880122-Summs, Jan. 1988.

[65]Cuta, J. M. and Krotiuk, W. J., "Reduced Gravity Boiling and Condensing Experiments Simulated with the COBRA/TRAC Computer Code," AIAA Paper 88-3634, July 1988.

[66]Dukler, A. E., Fabre, J. A., McQuillen, J. B., and Vernon, R., "Gas Liquid Flow at Microgravity Conditions: Flow Patterns and Their Transitions," *International Symposium on Thermal Problems in Space Based Systems*, Vol. 83, American Society of Mechanical Engineers, New York, Dec. 1987.

[67]Krotiuk, W. J., "Low Gravity Fluid Thermal Experimentation," presented at the Third Pathways to Space Experimental Workshop, Orlando, FL, June 1987.

[68]Cuta, J. M., Michener, T., Antoniak, Z. I., Bates, J. M., and Krotiuk, W. J., "Reduced Gravity Two Phase Flow Experiments in the NASA KC-135," *Transactions of Fifth Symposium on Space Power Systems*, CONF-880122-Summs, Jan. 1988.

[69]Chow, L. C. and Parish, R. C., "Condensation Heat Transfer in a Microgravity Environment," AIAA Paper 86-0068, Jan. 1986.

[70]Chow, L. C. and Parish, R. C., "Condensation Heat Transfer in a Microgravity Environment," *Journal of Thermophysics*, Vol. 2, Jan. 1988, pp. 82–84.

[71]Ambrose, J. H., Chow, L. C., Tilton, D. E., and Mahefkey, E. T., "Low Gravity Thermal Management Using Dielectrophoretic Forces," *Transactions of the Fifth Symposium on Space Nuclear Power Systems*, CONF-8801222-Summs, Jan. 1988.

[72]Lovell, T. K., "Liquid-Vapor Flow Regime Transitions for Use in the Design of Heat Transfer Loops in Spacecraft: An Investigation of Two Phase Flow in Zero Gravity Conditions," Air Force Wright Aeronautical Lab., OH, AFWAL TR-85-3021, May 1985.

[73]Theofanous, T. G., "Modelling of Aerospace (Microgravity) Systems," presented as Lecture 17 of the Workshop on Modelling of Two Phase Flow Systems, University of California, Santa Barbara, CA, Aug. 11–15, 1986.

[74]Hill, D. G., Hsu, K., Parish, R., and Dominick, J., "Reduced Gravity and Ground Testing of a Two-Phase Thermal Management System for Large Spacecraft," Society of Automotive Engineers, Paper 881084, July 1988.

[75]Hill, D. G. and Downing, R. S., "A Study of Two-Phase Flow in a Reduced Gravity Environment—Final Report," Sundstrand Energy Systems, Rockford, IL, Oct. 16, 1987.

[76]Chen, I., Downing, R. S., Parish, R., and Keshock, E., "A Reduced Gravity Flight Experiment: Observed Flow Regimes and Pressure Drops of Vapor and Liquid Flow in Adiabatic Piping," to be published.

[77]Giarratano, P. J., Owen, R. B., and Arp, V. D., "Transient Heat Transfer Studies in Low-Gravity Using Holographic Interferometry," AIAA Paper 88-0346, Jan. 1988.

[78]Yeh, T. P. and Orton, G. F., "Analytical and Experimental Modeling of Zero/Low Gravity Fluid Behavior," AIAA Paper 87-1865, June–July, 1987.

[79]Concus, P., "Equilibrium Fluid Interfaces in the Absence of Gravity," *Symposium on Microgravity Fluid Mechanics*, Vol. 42, American Society of Mechanical Engineers, New York, Dec. 1986, pp. 37–38.

[80]Napolitano, L. G., "Marangoni Convection in Space Microgravity Environments," *Science*, Vol. 225, July 13, 1984, pp. 197–198.

[81]Guyenne, T. D. (ed.), *Fifth European Symposium on Material Sciences Under Microgravity—Results of Spacelab-1*, European Space Agency, Paris, France, esa SP-222, Dec. 1984.

[82]Zell, M. and Straub, J., "Microgravity Pool Boiling—TEXUS and Parabolic Flight Experiments," *Sixth European Symposium on Material Sciences Under Microgravity Conditions*, European Space Agency, Paris, France, esa SP-256, Feb. 1987, pp. 155–160.

[83]Zell, M., Straub, J., and Weinzierl, A., "Nucleate Pool Boiling in Subcooled Liquid Under Microgravity—Results of TEXUS Experimental Investigations," *Fifth European Symposium on Material Sciences Under Microgravity*, European Space Agency, Paris, France, esa SP-222, Dec. 1984, pp. 327–333.

[84]Fuchs, M., Kretzschmar, K., and Stockfleth, H., "Microgravity with Balloon-Drop Capsules," *Sixth European Symposium on Material Sciences Under Microgravity Conditions*, European Space Agency, Paris, France, esa SP-256, Feb. 1987, pp. 459-462.

[85]Frimout, D. and Gonfalone, A., "Parabolic Aircraft Flights—An Effective Tool in Preparing Microgravity Experiments," esa Bulletin 42, May 1985, pp. 58–63.

[86]Gonfalone, A., "The Fluid Physics Module—A Technical Description," *Fifth European Symposium on Material Sciences Under Microgravity*, European Space Agency, Paris, France, esa SP-222, Dec. 1984, pp. 3–7.

[87]Napolitano, L. G., Monti, R., and Russo, G., "Some Results of the Marangoni Free Convection Experiment," *Fifth European Symposium on Material Sciences Under Microgravity*, European Space Agency, Paris, France, esa SP-222, Dec. 1984, pp. 15–22.

[88]Chun, C. H., "Verification of Turbulence Developing from the Oscillatory Marangoni Convection in a Liquid Column," *Fifth European Symposium on Materials Sciences Under Microgravity*, European Space Agency, Paris, France, esa SP-222, Dec. 1984, pp. 271–280.

[89]Schwabe, D. and Scharmann, A., "Measurements of the Critical Marangoni Number of the Laminar ↔ Oscillatory Transition of Thermocapillary Convection in Floating Zones," *Fifth European Symposium on Material Sciences Under Microgravity*, European Space Agency, Paris, France, esa SP-222, Dec. 1984, pp. 281–289.

[90]Napolitano, L. G., Monti, R., Russo, G., and Golia, C., "Comparison Between D1-Spaceborne Experiment and Numerical/Ground-Experimental Work on Marangoni Flows," *Sixth European Symposium on Material Sciences Under Microgravity Conditions*, European Space Agency, Paris, France, esa SP-256, Feb. 1987, pp. 191–199.

[91]Limbourg, M. C., Legros, J. C., and Petre, G., "Marangoni Convection Induced in a Fluid Presenting a Surface Tension Minimum as a Function of the Temperature," *Sixth European Symposium on Material Sciences Under Microgravity Conditions*, European Space Agency, Paris, France, esa SP-256, Feb. 1987, pp. 245–249.

[92]Neuhaus, D. and Feuerbacher, B., "Bubble Motions Induced by a Temperature Gradient," *Sixth European Symposium on Material Sciences Under Microgravity Conditions*, European Space Agency, Paris, France, esa SP-256, Feb. 1987, pp. 241–244.

[93]Padday, J. F., "The Behavior and Management of Liquid Systems in Low Gravity," *Sixth European Symposium on Material Sciences Under Microgravity Conditions*, European Space Agency, Paris, France, esa SP-256, Feb. 1987, pp. 49–55.

[94]Langbein, D., "Liquids and Interfaces," *Fifth European Symposium on Material Sciences Under Microgravity*, European Space Agency, Paris, France, esa SP-222, Dec. 1984, pp. 443–447.

[95]Halpern, R. E., "Facilities and Programs for Research in a Low-Gravity Environment," *Opportunities for Academic Research in a Low-Gravity Environment, Progress in Astronautics and Aeronautics*, series edited by G. A. Hazelrigg and J. M. Reynolds, Vol. 108, AIAA, New York, 1986.

[96]Vernon, R. W., private communication, NASA Lewis Research Center, Cleveland, OH, Nov. 21, 1985.

[97]Petrash, D. A. and Corpas, E. L., "Zero Gravity Facility for Space Vehicle Fluid Systems Research," *Proceeding of the 19th Annual Meeting of the Institute of Environmental Sciences*, Institute of Environmental Science, Mt. Prospect, IL, 1973.

[98]"Industry Observer," *Aviation Week and Space Technology*, May 16, 1988, p. 11.

[99]Williams, R., Houston, Texas, private communication, NASA Johnson Space Center, March 6, 1986.

[100]Shurney, R. E. (ed.), "The Marshall Space Flight Center KC-135 Zero Gravity Test Program for FY 1981," NASA TM-82476, Huntsville, AL, March 1982.

[101]Sears, F. W. and Zemansky, M. W., *University Physics*, 2nd ed., Addison-Wesley, Reading, MA, 1962, Chap. 34.

[102]Anderson, E. P., *Electric Motors*, 3rd ed., Audel, Indianapolis, IN, 1977, pp. 7–34.

[103]Murray, R. L., *Introduction to Nuclear Engineering*, 2nd ed., Prentice-Hall, Englewood Cliffs, NJ, 1961, pp. 204–207.

[104]Shawhan, S. D., "A Spacelab Principal Investigator's Guidance for Planning Scientific Experiments Using the Shuttle," *Journal of Spacecraft*, Vol. 20, Sept.-Oct. 1983, pp. 477–483.

[105]Salzman, J. A., "Two Phase Fluid Management Technology Base," presented at the NASA In-Space Research, Technology, and Engineering Workshop, Williamsburg, VA, Oct. 8–10, 1985.

[106]Hamacher, H., Jilg, R., and Merbold, U., "Analysis of Microgravity Measurements Performed During D1," *Sixth European Symposium on Material Sciences Under Microgravity Conditions*, European Space Agency, Paris, France, esa SP-256, Feb. 1987, pp. 413–419.

[107]Shelley, C. D., "Space Station Customer Utilization," presented at the National Aeronautics and Space Administration (NASA) In-Space Research, Technology, and Engineering Workshop, Williamsburg, VA, Oct. 8–10, 1985.

Chapter 5. Isothermal Gas-Liquid Flow
at Reduced Gravity

Isothermal Gas-Liquid Flow at Reduced Gravity

A. E. Dukler*

University of Houston, Houston, Texas

Nomenclature

$C_0 = U_b / U_{LS} + U_{GS}$

$\langle d_B \rangle$ = average bubble diameter

D = tube diameter

f_g = friction factor for gas phase flowing alone

f_i = interfacial friction factor

f_w = wall friction factor

$\langle L_S \rangle$ = average slug length

$\langle L_{TB} \rangle$ = average length of Taylor bubble

P = pressure

U_b = bubble velocity in slug flow

U_G = mean gas velocity

U_{GS} = superficial velocity of the gas (calculated as if it flowed alone in the pipe)

U_L = mean liquid velocity

U_{LS} = superficial velocity of the liquid (calculated as if it flowed alone in the pipe)

$U_M = U_{LS} + U_{GS}$

U_s = velocity of the liquid slug

$\langle U_w \rangle$ = average wave velocity

Copyright © 1989 by the American Institute of Aeronautics and Astronautics, Inc. All rights reserved.

*Professor, Chemical Engineering Department.

Aspects of this chapter dealing with flow patterns are based in part on material entitled "Gas Flow at Microgravity Conditions: Flow Patterns and Their Transitions," published in *International Journal of Multiphase Flow*, Vol. 14, 1988, pp. 389–400, with permission.

Re_{GS} = Reynolds number of the gas calculated as if it flowed alone in the pipe

Re_{LS} = Reynolds number of the liquid calculated as if it flowed alone in the pipe

z = axial distance

β = ratio of the slug length to the length of slug unit

ϵ = fraction voids

λ = average wavelength for annular waves

ρ_G = gas density

ρ_L = liquid density

τ_i = interfacial shear stress

τ_w = wall shear stress

$\langle \omega \rangle$ = average slug frequency

Introduction

Two-phase gas-liquid flow is expected to exist for a wide variety of applications in space. In addition to power cycles, propulsion, and thermal management systems, multiphase operations will be required for a wide variety of manufacturing processes involving condensation and boiling for separations and chemical reactions.

The reliable design of equipment for these purposes or the prediction of performance characteristics of specific equipment will require models that give a detailed description of the mechanics of such flows. Of most general interest is the manner in which the phases distribute (the flow patterns) and the pressure drop and its fluctuation. But there are situations for which much more detailed information may be necessary, including the size and velocity of the liquid slugs if they exist as well as the forces generated on the confining equipment and their characteristic frequency. When bubbles or drops are present, it is frequently required to have information about their size and velocity, including an analysis of the heat and mass transfer accompanying the flow. If annular flow takes place, of importance are models for the interfacial shear, the interfacial wave structure, the velocity distribution in the two phases, the radial distribution of bubbles or drops, and an understanding of the process of droplet creation from the continuous liquid phase.

The existence of reliable models based on good physical understanding of the flow mechanics is particularly important for space applications since the luxury of building empirical design correlations based on experiment would be prohibitively expensive if these experiments were to be carried out in space. Space applications of interest will undoubtedly involve the presence of two-phase flow along with condensation or boiling under nonadiabatic conditions. Paralleling recent successful research strategies at 1 g, we build the basic models for the simpler case of adiabatic flows without phase transfer and then expect to extend the models to the more complex case of heat and mass transfer between phases. Similarly, models can be developed for fractional gravity and then a methodology developed for predicting the unsteady behavior as a system enters or leaves the low-gravity environment.

During the past 10 yr there has been an extraordinary burst of attention to modeling gas-liquid flow under normal-gravity conditions. Physically based models now exist or will exist shortly for many of the situations previously described. However, on Earth the force due to gravity plays a dominant role in controlling the behavior of two-phase systems. For example, over a fairly wide range of flow rate space the flow pattern of gas-liquid flow in a horizontal tube can change drastically when the pipe is inclined upward as little as 1 deg from the horizontal.[1] Even at this small inclination, the component of the force of gravity acting in the axial direction exceeds the force due to wall shear stress. Clearly this condition will not exist at reduced gravity.

Calculations show that in larger conduits (above about 2.5 cm diam) at 1 *g* the forces along the free surfaces are usually small compared to the inertial and gravity forces that act. This has been confirmed by numerous studies that show a weak influence of interfacial tension on pressure drop, holdup, and flow patterns. For microgravity this condition can also be expected to be different.

Thus, it becomes necessary to reexamine existing models that describe the character of gas-liquid flow on Earth and to modify or construct new ones to provide valid descriptions of what happens at microgravity. Essential to a process as complex as this one is the acquisition of reliable experimental data to provide insight into mechanisms on which physical models can be based and which can be used to test the results.

This chapter summarizes the results of the initial engineering research studies of adiabatic gas-liquid flow under reduced gravity. The research was carried out under the sponsorship of the NASA Lewis Research Center. Experimental data were obtained in drop tower tests and in the LeRC Learjet.

Previous Studies

A considerable amount of experimental data now exist on reduced-gravity pool boiling, and this is discussed in another chapter of this volume. In contrast, the published research on either experiments or modeling for the *flow* of two phases is sparse. Albers and Macosko[2,3] measured pressure drop of condensing mercury at 0 *g*, and Namkoong et al.[4] made a photographic study of the 0-*g* mercury condensation process. Williams et al.[5] reported on the condensation of R-12 in a tube of 2.62 mm × 1.83 mm long at 1 *g* and showed a single KC-135 aircraft trajectory experiment. Qualitative comparisons were made for the two conditions.

The most extensive work on 0-*g* gas-liquid flow was presented by Hepner et al.[6,7] KC-135 trajectories were used to collect data on flow patterns for air-water flow in a 2.54-cm-diam tube having a length-to-diameter ratio (*L/D*) of 20. Although the test section length was short and duplicate tests gave significantly different results, the work is a landmark study. Unfortunately, the original motion picture films apparently are no longer available. Over the past 10 yr improved methods have been developed to interpret high-speed films taken of two-phase flow. Thus, if available, it is likely that these films would be interpreted differently today.

There have recently been a variety of conceptual studies on flow pattern transitions in space usually connected with proposed hardware design for space experiments. However, these offer little in the way of new insights. Lovell[8] attempted to construct an experimental analog of 0-g two-phase flow by using two fluids of near-identical density (water and polypropylene glycol). Experiments were carried out in a glass tube 2.54 cm diam × 6.3 m long. Serious questions must be raised as to the validity of the simulation.

Experimental Equipment

Experiments were carried out both on the Lewis 100-ft drop tower and on the Lewis Learjet using water and air. The data from both experimental flow loops consisted of the flow rates, temperatures, movie films taken at about 400 frames/s along with time-dependent pressure drop data. However, a much greater level of detail during each run was obtained in the Learjet tests because it was possible to construct a larger experimental rig equipped with a more complete data acquisition system. The drop tower provided about 2.2 s of near 0 g, whereas with the Learjet it was possible to collect data over 12–22 s depending on the quality of the particular trajectory. Accelerometer measurements, taken with each run, made it possible to limit data acquisition on the Learjet to periods when the acceleration did not exceed ±0.02 g in any of the three principal coordinates of the plane.

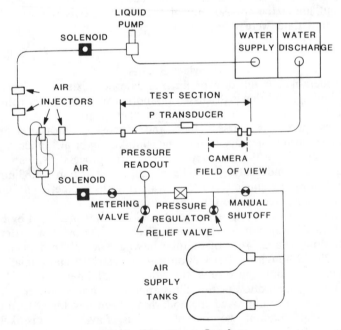

Fig. 1   Drop tower flow loop.

A schematic diagram of the drop tower loop appears in Fig. 1. The test section consisted of a transparent Plexiglas tube 9.52 mm in diameter and 0.457 m long which was backlit. At the mixer, air was injected into the liquid through four peripheral holes. Flow rates were set while the rig was suspended on its platform at the top of the tower. A calibrated valve was used for regulating air rate while the speed of a voltage controlled centrifugal pump regulated the liquid flow rate. A film record was taken at this 1-*g* condition. Then without changes in the settings the drop was executed with the camera activated. The rig was not equipped with flowmeters, and it was assumed that the pump speed and flow through the air control valve would not change during the drop. Subsequent analysis indicated that, due to drawdown of the batteries during the drop, the pump speed did change for certain runs. The pressure drop measuring system was not equipped with water flushing for the lead lines, and the data thus collected were not considered reliable enough to use in the analysis.

The test loop for the Learjet, whose schematic diagram appears in Fig. 2, was designed to overcome many of the limitations revealed from the drop tower experience. Air flow was metered through two critical flow orifices (one for high and one for low flow rates) with liquid measured using a turbine meter. Reading of the flow and temperature transducers were taken continually at 1.55-s intervals during the run. The test section pressure taps were equipped with reverse flush circuits that were activated just before the trajectory reached 0 *g*. Varian pressure transducers were

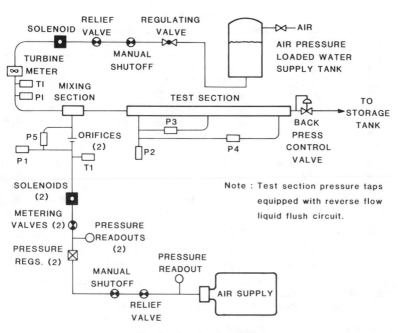

Fig. 2 Learjet flow loop.

used for measuring these pressure drops. Test section pressure drop data were collected at intervals of 0.002 s. Total straight length of the test section was 1.06 m with a diameter of 12.7 mm. The transparent test section was backed by a meter stick marked in millimeters and the rig was equipped with an LED display that indicated elapsed time in 0.01-s intervals. Color films were taken at 400 frames/s. Thus, in addition to flow pattern indications, it was possible to obtain measurements of velocities of slugs, bubble, and interfacial waves as well as sizes of bubble and slugs. In these measurements corrections for parallax errors were made. For each run at normal gravity a corresponding run was executed at 0 g at the same gas and liquid rate after the rig had been removed from the Learjet.

A study of the films showed that the flow patterns and the other characteristics of the flow were unchanged after the first 1.5 s into microgravity conditions. The calculated passage time for a continuity wave to traverse the test section was never greater than 1.2 s and is thus consistent with this determination.

Qualitative Observations

Table 1 summarizes the flow conditions for the tests that were carried out in the drop tower, and Table 2 shows those executed in the Learjet. Included is the flow pattern as deduced from a study of the movie films. Bubbly flow is designated when the gas bubbles are of a size less than or

Table 1　Drop tower data

Run no.	U_{GS}, m/s	U_{LS}, m/s	Flow pattern	Re_{LS}	Re_{GS}
1	0.252	0.657	Bubble: 1D B[a]	262	8344
2	0.252	0.444	Bubble: 1D B[a]	262	5639
3	0.252	0.278	Slug: 2D-4D TB[a]	262	3531
4	0.421	0.562	Bubble: 1D B[a]	438	7137
5	0.421	0.369	Slug: short slugs	438	4686
6	0.421	0.171	Slug: short slugs	438	2172
7	0.230	0.950	Bubble	240	12,065
8	0.230	0.749	Bubble	240	9512
9	0.230	0.532	Bubble	240	6756
10	0.230	0.250	Bubble	240	3175
12	0.460	0.697	Bubble	479	8852
13	0.460	0.442	Bubble	479	5613
14	0.460	0.175	Slug: 1D TB[d]	479	2223
16	0.690	0.598	Slug: 2D TB[e]	719	7595
17	0.690	0.366	Bubble: 1 D B[a]	719	4648
18	0.690	0.142	Slug: 2D-3D TB[b]	719	1803

[a]1D B: Bubble about 1 diam long.
[b]2D-3D TB: Taylor bubbles of length varying from 2 to 4 diam, etc.
[c]2D-4D TB: Taylor bubbles of length varying from 2 to 4 diam, etc.
[d]1D TB: Taylor bubble about 1 diam long.
[e]2D TB: Taylor bubble about 2 diam long.

equal to the tube diameter. The term "slug flow" is used when there are gas bubbles greater in length than the tube diameter and where there are regions along the tube where the liquid completely covers the flow area of the tube, even though this liquid may carry dispersed gas bubbles. Annular flow is the condition where the liquid never bridges the tube.

The information obtained from a study of the movie films is arrived at, in part, by observing a sequence of successive frames. However, in order to convey at least part of the impression, a series of black and white stills have been prepared and are presented in this section. Since these black and white photographs are very inferior copies of the original color pictures and only represent a single observation over a 1/400-s interval, hand traces of a different frame for each run have also been prepared to illustrate some details and the diversity of appearance.

Figure 3a shows photographs of single frames taken from the films for four gas rates at a low liquid rate of about 0.08 m/s. Hand traces of different frames in the same runs appear in Fig. 3b. Flow is from left to right. At low gas rate (run 15.1), well-established, stable, spherically nosed "Taylor" bubbles that are axisymmetric move along the pipe separated by clear liquid slugs. The back of the bubbles generally assume a shape suggested by Coney and Masica.[9] Bubble and slug lengths vary, but the variance is relatively small. As the gas rate is increased (test 2.2), the bubbles become longer and in some cases carry very thin membranes that bridge the bubble. Both the bubbles and slugs have much greater variance in length, and the slugs contain some gas in the form of dispersed smaller bubbles. At still higher gas rates as in run 11.1, a condition very close to transition between slug and annular flow appears. Long stretches of nearly smooth film are occasionally disrupted by a slow-moving high-amplitude wave that sometimes is seen to bridge the pipe forming a small liquid slug. In many cases these slugs do not persist, breaking up into a locally thick annular film. Away from the point of slug inception the film is remarkably smooth. At a high gas rate (run 17.1) the film is very wavy with occasional large roll waves sweeping by at velocities approaching the gas velocity.

Figures 4a and b show conditions that exist at a much higher liquid rate, approaching 1 m/s. At the lowest gas rates (run 15.2), the gas is seen to be dispersed in the liquid in the form of bubbles from 0.2 to 0.5 mm in characteristic dimension. At higher gas rates (run 14.1), the bubbles become smaller and more closely packed but are still dispersed with what appears to be relative uniformity in the axial direction. However, the next run in this series (run 12.12) displays a high flow rate slug flow. A rapidly moving highly aerated liquid slug is separated from the next slug by a thick, wavy liquid film that itself carries bubbles. The front and back of the slug are clearly defined even though the degree of aeration is high.

A condition of intermediate gas and liquid rate is shown in Fig. 5. Here the slug is substantially aerated, but the bubbles separating the slugs are regular and their films are free of gas bubbles.

The films and sketches show that these visual observations can provide a great deal of information by which to construct and test simple models for the flow. Conditions of transition can be estimated, the velocity of the slugs

Table 2   **Learjet data**

Run no.	U_{GS}, m/s	U_{LS}, m/s	Flow pattern[a]	Re_{LS}	Re_{GS}
2.1	25.32	0.080	Annular: short roll waves	1016	26,368
2.2	0.61	0.084	Slug: long TB	1067	635
3.1	11.44	0.451	Annular: roll waves	5728	11,914
4.1	7.97	0.082	Annular: roll waves	1041	8300
4.2	0.22	0.076	Slug: 4D TB	965	229
5.1	2.22	0.079	Annular: near transition	1003	2312
5.2	0.64	0.080	Slug: long TB	1016	666
7.1	2.99	0.438	Annular: near transition	5563	3114
8.1	1.09	0.460	Slug: 3D TB, bubbly slugs	5842	1135
9.1	0.09	0.478	Bubble	6071	94
10.1	23.00	0.418	Annular: roll waves	5309	23,952
11.1	1.80	0.079	Transition: annular-slug	1003	1875
11.2	1.75	0.450	Slug: long TB, bubbly slugs	5715	1822
12.1	1.90	0.920	Slug: 6D TB, bubbly slugs	11,684	1979
13.1	0.70	0.080	Slug: long TB, nonbubbly slugs	1016	729
13.2	0.65	0.450	Slug: 2D TB, bubbly slugs	5715	677
14.1	0.65	0.940	Bubble	11,938	677

(Table 2 continued on next page.)

Table 2 (cont.) Learjet data

Run no.	U_{GS}, m/s	U_{LS}, m/s	Flow pattern[a]	Re_{LS}	Re_{GS}
15.1	0.16	0.079	Slug: 2D 3D TB, nonbubbly slugs	1003	167
15.2	0.13	0.880	Bubble	11,176	135
16.1	11.40	0.077	Annular: roll waves	978	11,872
16.2	0.13	0.460	Bubble	5842	140
17.1	10.10	0.080	Annular: roll waves	1016	10,518
24.1	1.92	0.07	Slug: long TB	889	1999
24.2	10.5	0.484	Annular	6147	10,935
25.1	11.22	0.073	Annular	927	11,685
25.2	0.68	0.79	Bubble	10,033	708
26.1	0.817	0.0704	Slug: long TB	894	851
26.2	0.774	0.454	Bubble: near transition	5766	806
27.1	2.25	0.208	Annular: near transition	2642	2343
27.2	0.219	0.21	Slug: short TB	2667	228
28.1	0.215	0.446	Bubble	5664	224
28.2	10.1	0.2	Annular	2540	10,518
29.1	0.1	0.07	Slug: short TB	889	104
29.2	4.17	0.46	Annular	5842	4343
30.1	11.1	0.187	Annular	2375	11,560
30.2	0.7	0.575	Bubble	7303	729

[a]See Table 1 footnote for explanation of comments in flow pattern column.

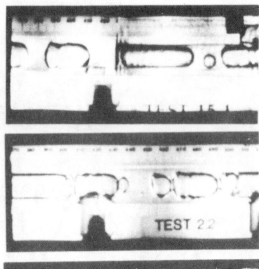

Fig. 3a) Single frames from four Learjet runs: $U_{LS} \simeq 0.008$ m/s.

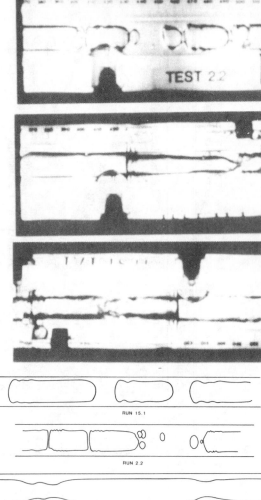

Fig. 3b) Sketches of movie frames.

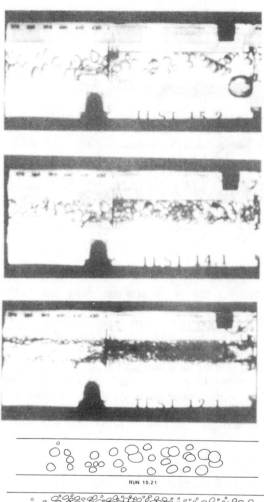

Fig. 4a) Single frames from three Learjet runs: $U_{LS} \simeq 0.9$ m/s.

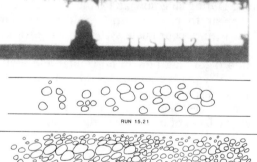

Fig. 4b) Sketches of movie frames.

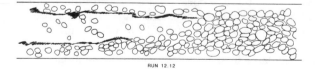

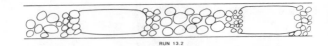

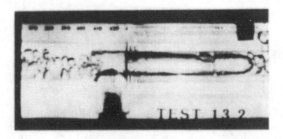

Fig. 5 Sketch and single frame from a Learjet run (intermediate gas and liquid rates): $U_{GS} \simeq 0.65$ m/s **and** $U_{LS} \simeq 0.45$ m/s.

and bubbles can be calculated, and the variation of these velocities with position of the dispersed bubbles can be determined. Slug and bubble lengths can be measured for slug flow, and some estimate of bubble size can be evolved for the distributed bubbly pattern. Thus, these flow visualization experiments carry with them much information, and this type of analysis is under way at this time. It is already possible to indicate one unexpected result from these observations. During slug flow at 1 g the bubbles carried in the liquid slugs always appear to have a "drift" velocity measured relative to the slug itself. That is, in a coordinate system moving with the slugs the dispersed bubbles appear to move backward. This observation is consistent with models developed for slug flow in horizontal and vertical pipes by Dukler and Hubbard[10] and Fernandes et al.[11] However, at reduced-gravity conditions these dispersed bubbles move at precisely the velocity of the front of the slug, suggesting that the mechanism of pickup and shedding that has formed the basis of modeling in the past may not be applicable here.

Flow Pattern Mapping

A map of the flow patterns observed appears in Fig. 6. Data for both the drop tower and the Learjet are included. In general, tube diameter can be expected to have an effect on the location of the transition boundaries in these coordinates of superficial velocity, U_{LS} and U_{GS}. The superficial velocity is that velocity calculated as if the phase were flowing alone in the pipe. However, the two test section diameters are not drastically different. Furthermore, the drop tower data include only the patterns of bubbly and slug flow, and models show that the transition between bubbly and slug flow is relatively insensitive to diameter.

At this time there is a question as to whether the bubble and slug flow regions should be considered separate patterns or whether this series of

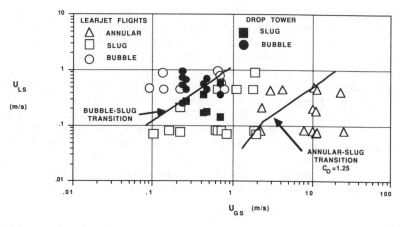

Fig. 6 Flow pattern map for reduced gravity and comparison with transition models.

runs simply represents a continuum of bubble sizes. At 1 *g* physical models have been developed which suggest that the mechanism by which the flow takes place changes drastically between these two patterns. However, preliminary analysis for microgravity indicates that these two regions may represent a continuum of the same physical process. If that proves to be the case, only two patterns can be considered to characterize the flow: bubbly and annular.

Modeling of the flow pattern transitions is in its earliest stages; however, it is possible to suggest some simple ideas by which the location of transition boundaries can be estimated.

Bubble to Slug Pattern

Study of the movie films clearly indicates that the local slip between liquid and gas is negligible. Thus, one can write

$$U_L = U_G \tag{1}$$

where these are the space-averaged true linear velocities of the liquid and gas, respectively. Designate ϵ as the area average void fraction. The linear velocities and the superficial velocities are related by

$$U_L = U_{LS}/(1 - \epsilon), \qquad U_G = U_{GS}/\epsilon \tag{2}$$

where the superficial velocities are computed as if that phase were flowing alone in the tube. Substituting gives

$$U_{LS}/U_{GS} = (1 - \epsilon)/\epsilon \tag{3}$$

The transition from bubble to slug flow is thought to take place when the bubble concentration and size are such that adjacent bubbles come into contact. Then coalescence can be expected, and surface tension causes the two coalescing bubbles to form one larger characteristic of slug flow. Thus, one needs only to estimate the average voids at this condition to obtain an equation relating the superficial velocities at transition. Small bubbles in a cubic array can achieve, at most, a void fraction of 0.52. However, large bubbles, approaching that of the diameter of the tube, will generate a holdup before touching that depends on their shape and orientation. For large spherical bubbles this can be shown to be approximately at $\epsilon = 0.5$. However, for ellipsoids the void fraction will depend on whether the major axis is aligned with the axis of the tube or with the radius when the voids will approximate 0.4. Because one observes various alignments, it is speculated that the average void fraction at contact (and thus at transition) is approximately $\epsilon = 0.45$. The resulting equation is then

$$U_{LS} = 1.22 U_{GS} \tag{4}$$

This equation is plotted in Fig. 6 and appears in reasonable agreement with the transition shown by the data. The result is equally satisfactory if $\epsilon = 0.5$ is used. Note that according to this model the transition in coordinates of superficial velocity is independent of diameter.

Slug to Annular Pattern

The following mechanism is hypothesized to take place causing this transition. During slug flow there is a large axial variation in void fraction between the slugs and the Taylor bubbles. As the gas rate is increased the length of the bubbles increases relative to the slug lengths. When these slugs become short enough, slight variation in local velocity or adjacent film thickness can cause the slug to momentarily rupture. Then surface tension forces draw the liquid around the wall of the pipe to establish annular flow and the slug cannot be reformed. In order to estimate the flow conditions at which this change will take place, equations are developed relating the axial average voids and the superficial flow rates for slug flow. A similar relation is developed for annular flow. It is speculated that the transition between slug and annular flow takes place when the void fraction as dictated by the two models first becomes equal. That is, at lower gas velocities, the slug flow model always predicts higher average voids than does the annular flow model at the same flow rates. However, at the transition velocity the voids predicted by the two models are equal. At still higher gas flow rates the slug flow model predicts lower voids than does the annular flow model; thus, the flow pattern becomes one of annular flow since surface tension will cause the liquid to wrap around the wall instead of existing in discrete slugs.

A model for the *average voids in slug flow* can be approached as follows: Consider a typical slug unit consisting of one Taylor bubble of length l_b moving at a velocity U_b and its adjacent slug having corresponding quantities l_s and U_s. A material balance on the gas gives

$$U_{GS} = U_s \epsilon_s \beta + U_b \epsilon_b (1 - \beta) \tag{5}$$

where $\beta = l_s/(l_s + l_b)$. As discussed earlier, the flow visualization shows that the slug and bubble velocities are equal. Thus, the material balance simplifies to

$$U_{GS}/U_b = \epsilon_s\beta + \epsilon_b(1 - \beta) = \langle\epsilon\rangle \qquad (6)$$

where $\langle\epsilon\rangle$ is the average void fraction during slug flow. Modeling of slug flow at microgravity is now under way and should produce a basis for predicting the ratio of the gas superficial velocity to the Taylor bubble velocity. Designate

$$C_0 = U_b/(U_{LS} + U_{GS})$$

which is recognized to be a measure of the rate at which the large bubble advances ahead of the two-phase mixture in slug flow and is a basic parameter of slug flow modeling. Then, substituting into Eq. (6) gives

$$U_{GS}/(U_{LS} + U_{GS}) = C_0\langle\epsilon\rangle \qquad (7)$$

Studies of the films show that C_0 ranges between 1.15 and 1.30, depending on the flow rates of the phases. In other systems this ratio may also depend on the fluid properties and pipe size. Until these slug flow modeling studies are completed, the average value is assumed to be 1.25 based on analysis of the experiments. Substituting into Eq. (6) then provides a relationship between the superficial velocities and average voids in slug flow.

Now consider the condition of *annular flow* where all of the liquid flows as a smooth film along the wall and the gas flows in the core. A force balance on a control volume bounded by the pipe walls and two planes normal to the axis separated by an axial distance Z gives

$$(P/Z) = 4\tau_w/D \qquad (8)$$

The force balance taken over the liquid film in that same control volume is

$$(P/Z)(1 - \epsilon) = (4\tau_w/D) - (4\tau_i/D)\epsilon^{\frac{1}{2}} \qquad (9)$$

In these equations P is the pressure gradient, τ_w the wall shear stress, and τ_i the interfacial stress. The pressure gradients in Eqs. (8) and (9) are equated to give

$$\tau_i = \tau_w\epsilon^{\frac{1}{2}} \qquad (10)$$

These stresses can be written in friction factor formulation:

$$\tau_i = (f_i\rho_G U_G^2)/2, \qquad \tau_w = (f_w\rho_L U_L^2)/2 \qquad (11)$$

Since

$$U_L = U_{LS}/(1 - \epsilon)$$

and

$$U_G = U_{GS}/\epsilon$$

$$\epsilon^{\frac{5}{2}}/(1 - \epsilon)^2 = (f_i/f_w)(\rho_G/\rho_L)(U_{GS}/U_{LS})^2 \tag{12}$$

Now it remains only to evaluate each friction factor in terms of the Reynolds number for that phase to arrive at an expression that gives the average voids in annular flow given the superficial velocities. The Blasius equation for a smooth surface is used for the wall:

$$f_w = C/(Re_L)^n \tag{13}$$

where $C = 16$ when the flow is laminar and 0.046 if turbulent and $n = 1.0$ for laminar flow and 0.2 if turbulent. Little is known at this time about the factors that determine the interfacial friction factor f_i, except that at the same gas Reynolds number it is much larger than for flow over a smooth rigid surface. The existence of interfacial waves is known to be the primary cause for this increase, although the precise mechanism of wave action is not yet understood. Experiments on Earth show that the details of the wave structure depend on the direction of gravity relative to the direction of flow of the thin film. For purposes of developing this model further at this time, a preliminary estimate of f_i is made using the empirical correlation suggested by Wallis[12]:

$$f_i/f_g = 1 + 150(1 - \epsilon^{\frac{1}{2}}) \tag{14}$$

where f_g is the single-phase friction factor calculated from a relationship similar to Eq. (13) with Re_G replacing Re_L.

Now it is possible to evolve the model for transition based on the mechanism suggested earlier. Equating $\langle \epsilon \rangle$ from Eq. (7) with ϵ of Eq. (12) provides the intersection between the models for slug flow and annular flow, at which point both models predict the same void fraction. This process, which eliminates ϵ between the two equations, provides a method by which the superficial velocities can be calculated at this transition condition. The result, calculated numerically for the measured value of $C_0 = 1.25$, is shown in Fig. 6. The discontinuity in the theoretical curve comes as a result of the transition between laminar and turbulent flow. Agreement is seen to be reasonably satisfactory.

A study has recently been completed by Sundstrand of two-phase flow of Freon-114 at 5.8 bars in a transparent pipe of 15.9 mm i.d. during KC-135 low-gravity trajectories carried out by the NASA Johnson Space Center. Details of the experiments are available,[13] and these properties are drastically different from those of the experiments conducted here. These movie films were provided by NASA and have now been analyzed. Only annular and slug flow patterns were observed. Measurements of the bubble velocities during slug flow showed that $C_0 = 1.06$ for these tests. Theoretical transition boundaries were calculated for this condition and are compared

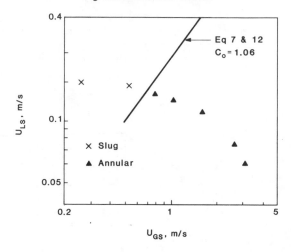

Fig. 7   Flow patterns for Sunstrand data: comparison with transition model.

with the Sundstrand data in Fig. 7 where very satisfactory agreement is observed.

It is important to recognize that these results represent preliminary efforts in this transition modeling process. Completing the effort will require a more detailed understanding of slug flow to predict the value of the coefficient C_0, along with a basic method for evaluating the effect of interfacial waviness on interfacial shear.

Characteristics of Annular Flow

Diagnostic studies carried out for each of the flow patterns provides insights into the mechanisms controlling the behavior for each pattern of phase distribution. Some preliminary results for the annular pattern are now presented.

Measurements were made of the pressure drop between taps at fixed positions along the test section for Learjet runs. These can be compared with the predictions of the model presented earlier for the transition between annular and slug flow. Equation (8) is used to compute the pressure gradient from the wall shear τ_w. The wall shear is computed from the interfacial shear from Eq. (10) once the void fraction ϵ has been determined. The ϵ is found from the iterative solution of Eq. (12) once the flow rates and fluid properties are specified, and the interfacial shear can be found from Eq. (11). Thus, it is possible for each flow rate pair to compute a pressure gradient, and these values are compared with experimentally measured ones in Fig. 8. Agreement is shown to be surprisingly good in view of the uncertainty of these measurements. It should be noted that the Wallis correlation for interfacial friction factor [Eq. (14)] is largely empirical, having been developed for low-pressure air/water flows in small pipes. It is useful to know that the shift to low gravity does not seem to affect the

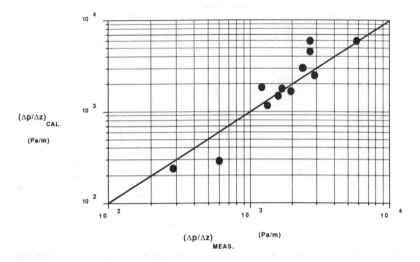

Fig. 8 Comparison of experimental and measured pressure gradient.

degree of agreement significantly, but the predictions remain to be tested at other fluid properties and pipe sizes for annular flow.

The characteristics of the interface are critical for building mathematical models to predict heat and mass transfer through the film. Dukler[14] has shown that the shape and velocity of the waves are the controlling factors. The films have been analyzed to obtain information on these characteristics and results are listed in Table 3. At high gas rates the wave length and velocity and thus the frequency are comparable to those observed at 1.0 g, as expected when the shear forces dominate the process. However, the wave lengths are significantly different than their 1.0-g analogs for low gas rates.

Table 3 Measured characteristics of the annular flow pattern

Run no.	U_{GS}, m/s	U_{LS}, m/s	$\langle U_w \rangle$, m/s	$\langle \omega \rangle$, 1/s	$\langle \lambda \rangle$, m
2.1	25.32	0.080	3.70	28	0.13
3.1	11.44	0.451	4.50	19	0.24
4.1	7.97	0.082	2.25	7	0.33
5.1	2.22	0.079	1.65	3	0.55
7.1	2.99	0.438	3.10	7	0.44
10.1	23.00	0.418	5.40	40	0.14
16.1	11.40	0.077	2.06	11	0.18
17.1	10.10	0.080	2.20	10	0.22
24.2	10.5	0.484	4.06	17	0.24
25.1	11.22	0.073	2.03	10	0.20
27.1	2.25	0.208	2.71	4	0.68
29.2	4.17	0.46	2.9	9	0.32
30.1	11.1	0.187	3.6	15	0.24

Table 4 Measured characteristics of the slug flow pattern

Run no.	U_{GS}, m/s	U_{LS}, m/s	U_M, m/s	$\langle L_{TB} \rangle$, m	$\langle L_S \rangle$, m	$\langle \omega \rangle$, 1/s
2.2	0.61	0.084	0.69	0.230	0.030	2.23
4.2	0.22	0.076	0.30	0.050	0.020	3.71
5.2	0.64	0.080	0.72	0.125	0.055	2.33
8.1	1.09	0.460	1.55	0.087	0.290	3.21
11.2	1.75	0.450	2.20	0.160	0.080	11.71
12.1	1.90	0.920	2.82	0.280	0.100	8.15
13.1	0.70	0.080	0.78	1.240	0.133	0.68
13.2	0.65	0.450	1.10	0.054	0.038	13.8
15.1	0.16	0.079	0.24	0.035	0.012	5.95
20.1	0.21	0.083	0.29	0.035	0.008	7.21
22.1	0.18	0.082	0.26	0.031	0.010	6.58
23.2	0.22	0.083	0.30	0.039	0.009	6.87
26.1	0.82	0.070	0.89	0.280	0.080	1.28
27.2	0.22	0.210	0.43	0.025	0.015	8.50
29.1	0.10	0.070	0.17	0.015	0.008	6.52

Characteristics of Slug Flow

All runs that displayed the slug flow pattern were analyzed to determine the mean value of the slug and bubble length and the slugging frequency. These data appear in Table 4 where quantities enclosed by $\langle \rangle$ represent sample means. The 30-fold variation in slug length is drastically different from that observed for 1.0-*g* systems and suggests that the interfacial tension plays a strong role in such processes.

Characteristics of Bubbly Flow

The bubble size that exists in a bubbly system controls the interfacial area available for heat and mass transfer. Measurements made of the mean bubble diameter are listed in Table 5.

Table 5 Measured characteristics of the bubbly flow pattern

Run no.	U_{LS}, m/s	U_{GS}, m/s	U_M, m/s	$\langle d_B \rangle$, cm
9.1	0.478	0.090	0.568	0.50
14.1	0.940	0.650	1.590	0.25
15.2	0.880	0.130	1.010	0.35
16.2	0.460	0.134	0.594	0.35
25.2	0.790	0.680	1.470	0.20
26.2	0.454	0.774	1.228	0.50
28.1	0.446	0.215	0.661	0.30
30.2	0.575	0.700	1.275	0.40

Summary of the Present State of the Art

A significant body of data now exists that provides insights into the mechanics of two-phase gas-liquid flow at reduced-gravity conditions. Preliminary models are in hand for the prediction of the flow pattern transition criterion in adiabatic systems for low viscosity and high interfacial tension systems. These have yet to be fully tested for other fluid properties and pipe sizes and must be extended to accommodate the processes of condensation and flow boiling. Based on experience at 1 g these extensions should offer no major difficulties.

The task at hand is to build comprehensive physical models for each of the flow patterns so that the characteristics of the flow can be predicted with confidence. Although a start on this process has been made and certain results are reported in this chapter, much remains to be accomplished before systems operating with two-phase flow in space can be designed with confidence.

Acknowledgment

The support of NASA under Grant NAG 3-510 to the University of Houston is gratefully acknowledged.

References

[1]Taitel, Y. and Dukler, A. E., "A Model for Predicting Flow Regime Transitions in Horizontal and Near Horizontal Gas-Liquid Flow," *AIChE Journal*, Vol. 22, 1976, pp. 47–55.

[2]Albers, J. A. and Macosko, R. P., "Condensing Pressure Drop of Non Wetting Mercury in a Uniformly Tapered Tube in 1G and Zero Gravity Environments," NASA TN D-3185, 1966.

[3]Albers, J. A. and Macosko, R. P., "Experimental Pressure Drop Investigation of Non Wetting, Condensing Flow of Mercury Vapor in a Constant Diameter Tube in 1-G and Zero Gravity Environments," NASA TN D-2838, 1965.

[4]Namkoong, D., Block, H. B., and Macosko, R. P., "Photographic Study of Condensing Mercury in 0 and 1 G Environments," NASA TN D-4023, 1967.

[5]Williams, J. L., Keshock, E. G., and Wiggins, C. L., "Development of a Direct Condensing Radiator for Use in a Spacecraft Vapor Compression Refrigeration System," *Journal of Engineering for Industry*, Vol. 96, 1973, pp. 1053–1063.

[6]Hepner, D. B., King, C. D., and Littles, J. W., "Aircraft Flight Testing of Fluids in Zero Gravity Experiments," General Dynamics Rept., CASD NAS-74-054, 1978.

[7]Hepner, D. B., King, C. D., and Littles, J. W., "Zero-G Experiments in Two-Phase Fluid Flow Regimes," American Society of Mechanical Engineers, Paper 75-ENAs-24, 1975.

[8]Lovell, T. K., "Liquid-Vapor Flow Regime Transitions for Use in the Design of Heat Transfer Loops in Spacecraft: An Investigation of Two Phase Flow in Zero Gravity Conditions," Air Force Wright Aeronautical Lab., Rept. TR-85-3021, 1985.

[9]Coney, T. A. and Masica, W. J., "Effect of Flow Rate on the Dynamic Contact Angle for Wetting Liquids," NASA TN D-5115, 1960.

[10]Dukler, A. E. and Hubbard, M. G., "A Model for Gas-Liquid Slug Flow in Horizontal and Near Horizontal Tubes," *Industrial and Engineering Chemistry Fundamentals*, Vol. 14, 1976, pp. 337–346.

[11] Fernandes, R., Semiat, R., and Dukler, A. E., "A Hydrodynamic Model for Gas-Liquid Slug Flow in Vertical Tubes," *AIChE Journal*, Vol. 29, pp. 981–989.

[12]Wallis, G. B., *One Dimensional Two Phase Flows*, McGraw-Hill, New York, 1969.

[13]Hill, D. and Downey, R. S., "A Study of Two Phase Flow in a Reduced Gravity Environment," Final Rept., Sundstrand Energy Systems, NASA Contract NAS9-17195, 1987.

[14]Dukler, A. E., "The Role of Waves in Two Phase Flow," *Chemical Engineering Education*, Vol. 11, 1977, pp. 108–119.

Chapter 6. Vapor Generation in Aerospace Applications

Low-Gravity Flow Boiling

Judith M. Cuta* and William J. Krotiuk†
Battelle, Pacific Northwest Laboratories, Richland, Washington

Large Power Level Devices

Power and other systems that use a working fluid to extract energy and provide useful work are characterized under the broad heading of thermal-hydraulic systems. Examples of such systems that might be used in space applications include nuclear reactors, solar power plants, environmental heating and cooling systems, and refrigeration systems for cryogenic fluid storage and handling. Because of the high cost of putting material in orbit and the limitations on size and weight imposed by the current space transportation system, systems deployed for power production in space need to be as efficient as possible. For thermal-hydraulic systems, this usually means making the working fluid undergo phase change as it cycles through the loop. The process takes advantage of the high heat-transfer rate in flow boiling, which is generally orders of magnitude more efficient than single-phase heat transfer.

Two-phase flow heat-transfer processes are well characterized for conditions on Earth, both in practical engineering terms and in theoretical understanding of the underlying physical phenomena. Design correlations for boilers, condensers, heat exchangers, etc., are readily available for normal-gravity application in nearly any possible two-phase flow system. Analytical techniques—some using relatively simple but reasonable approximations of the physics of two-phase flow, others using sophisticated

This work is declared a work of the U.S. Government and is not subject to copyright protection in the United States.

*Senior Development Engineer.

†Currently Principal Member Technical Staff, General Electric Astro-Space Division, Princeton, New Jersey.

This chapter is based in part on the material published in Cuta, J. M. and Krotiuk, W. J., "Reduced Gravity Boiling and Condensing Experiments Simulated with the COBRA/TRAC Computer Code," AIAA Paper 88-3634-CP, July 1988.

multicomponent representations of the heterogenous flowfield—have been developed for numerous computer simulation codes for two-phase flow and heat transfer under steady-state and severe transient conditions. But all of these approaches to the analysis and design of two-phase thermal-hydraulic systems require large amounts of experimental data to provide the under-pinnings of the design or analysis methods.

Correlations always contain empirical parameters that must be derived from data, no matter how sophisticated the physical modeling underlying their derivation. In some cases, the correlations are extremely specific, in that they give valid and reasonable results only over conditions that correspond exactly to their database and do not extrapolate reasonably in any direction. For more physically grounded correlations, their range of applicability is usually larger, but it is still of necessity limited. There is no currently available method of analyzing two-phase flow and heat transfer that proceeds purely from first principles. In even the most sophisticated analytical approach, reality intrudes in the form of experimental data to correlate empirical parameters required to achieve closure of the equation set characterizing the flow.

Correlations and constitutive models for two-phase flow and heat transfer derived in 1 g cannot, a priori, be assumed to apply in 0 g.[1,2] The gravity force, g, which is implicit in all terrestrial correlations as a fixed constant, is suddenly transformed into a fundamental variable in extraterrestrial applications. A whole new database is needed, derived from experiments over as wide a range of design conditions as possible, in order to accurately characterize the behavior of two-phase systems in reduced gravity. For purely economic reasons, these correlations need to be as accurate as possible, so that systems can be designed without wasted size and mass. They also need to be capable of performing consistently within design parameters, since the lives of people living and working in space will depend on the reliability and efficiency of these systems.

The existent data on two-phase flow behavior in reduced gravity is not sufficient to develop useful correlations and constitutive models for design of two-phase thermal-hydraulic systems in 0 g. (See Chapter 4 for a detailed summary of the current state of the low-gravity database.) The following subsections of this chapter discuss ways to address this lack of information. The following section contains a detailed presentation of one set of reduced-gravity experiments, discusses the differences from 1-g behavior uncovered, and indicates the significant questions raised that will require further investigation. The section on Needs for Two-Phase Thermal-Hydraulics in Reduced Gravity presents a discussion of future work needed to develop useful analytical and design tools for large-scale thermal-hydraulic systems in reduced gravity.

Reduced-Gravity Two-Phase Experiments

To support efforts to develop analytical design tools for application to space-based power systems, a series of reduced-gravity two-phase flow experiments were performed in the NASA KC-135 aircraft. These tests were conducted in conjunction with Dr. F. Best of Texas A&M University, using a test section built for NASA-sponsored low-gravity condensing studies.[3,4] Battelle, Pacific Northwest Laboratories (PNL) added a transparent boiler section with appropriate instrumentation to the test apparatus. The following sections discuss the experiment apparatus and results. Selected tests were simulated with the COBRA/TRAC computer code,[5] and the results are compared to test data. These comparisons illustrate the capabilities and limitations of existing analytical methods for predicting low-gravity fluid thermal behavior.

Reduced-gravity boiling and condensing experiments were flown on the NASA KC-135 aircraft. To permit the study of low-gravity boiling, the tests were performed in an instrumented boiler consisting of an 8-mm-i.d. quartz tube wrapped in a helix of nichrome wire. A schematic of the "once-through" test assembly[6] is shown in Fig. 1. Flow exiting the condenser was vented to the atmosphere outside the KC-135 aircraft. Pressure, temperature, heater power, flow rate, accelerometer, and gamma densitometer measurements were recorded. Measuring station locations are shown on the loop diagram in Fig. 2.

The test matrix for the flights consisted of 15 boiling/condensing tests: 12 unique tests, plus 3 repeat tests, to be run on the first day of testing in the KC-135. Ten nitrogen/water adiabatic tests were run on the second day. The planned and actual conditions for these tests are listed in Tables 1 and 2. The boiling/condensing test point flow rates would be expected to produce either stratified or slug-stratified flow for a horizontal orientation in 1 g. The adiabatic tests were primarily at low flow rates, where the horizontal 1 g flow regime would be slug, stratified slug, or bubbly flow.

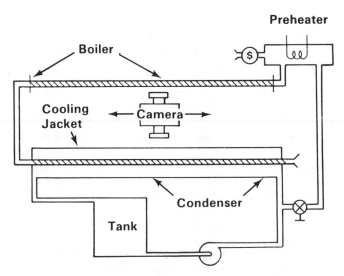

Fig. 1 Schematic of boiling/condensing KC-135 experiment apparatus.

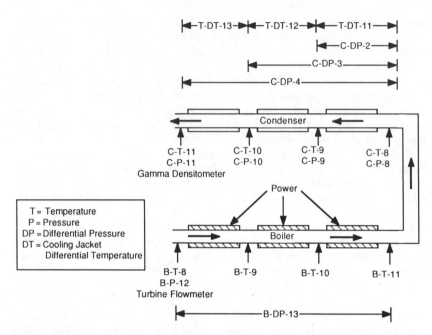

Fig. 2 Instrumentation measurement locations for boiling/condensing experiment.

The adiabatic tests were included in the full test matrix to obtain reliable two-phase data at the lower flow rates. During earlier tests, experience with the test section in boiling with very low flow rates had shown that the flow was very unstable, probably due to transition effects during a parabolic maneuver. The process of void formation in the boiler caused the flow to chug erratically and even reverse direction periodically. This tended to obscure and overwhelm the hydrodynamic behavior of the two-phase flow, making it difficult to interpret the significance of the data obtained at low flow rates.

In the KC-135, the test conditions were set up for a given test point in level flight, then data were taken during three successive 0-g parabolas. This permitted three 0-g periods of approximately 20 s for each test point, separated by a 1-min-long period of acceleraton at 1.7–1.9 g. The acceleration period ensured that the flow would be fully stratified before each transition to 0 g. For almost every test point, the flow regime changed to annular flow immediately upon entering 0 g. Data were taken continuously throughout the sequence of three parabolas for each test.

In addition to the instrument readings taken by the data acquisition system during the tests, recordings were made of the boiler and condenser flows using NASA high-speed cameras and a Kodak high-speed video imaging system. The NASA cameras operated at 2700 frames/s and provided approximately 4 s of real-time recording. The high-speed video system was provided on a demonstration basis by Eastman Kodak's Spin

Physics Division. It was operated at 1000 frames/s and was capable of recording up to 30 s of real time.

For the two-phase KC-135 tests to be acceptable, it must be ensured that the flow conditions reached a true steady state during the reduced-gravity interval. Also, the acceleration in the reduced-gravity interval must be low enough to represent 0-g conditions. These conditions were met in the boiler; the data show umambiguously that the fluid in the boiler was at thermal equilibrium (constant heat transfer) throughout the reduced-gravity interval, and the high-speed video images indicate a stable flow regime. In the condenser, however, there was some evidence that full thermal equilibrium was not achieved at the lowest flow rates until late in the reduced-gravity interval. This may have been due to sudden changes in the heat-transfer rate following the transition to reduced gravity, since the temperatures were constant prior to entering the reduced-gravity condition. The high-speed films obtained in the condenser show that the flowfield displayed no evolutionary changes, which indicates that momentum equilibrium was well established in the condenser in all tests. This is consistent with the observed behavior in adiabatic tests in drop towers and in the NASA Learjet.[7]

The reduced vertical acceleration during the low-gravity interval in the KC-135 is higher than that obtained in space. The accelerations are typically about $0.02 g$ in the KC-135, whereas in low Earth orbit the acceleration of gravity is on the order of $10^{-4} g$. The parabolic trajectory of the KC-135 results in the net vertical acceleration approximately following a parabolic curve. Figure 3 plots the analog signal from the vertical axis of the three-axis accelerometer in the KC-135 for parabola 10. It is typical of the parabolas flown on both days of testing.

The transitions into and out of reduced gravity take up about one-third of the total interval, leaving approximately 15 s of each parabola with a relatively steady low-gravity condition. Although conditions in the KC-135 only approximate the gravitational environment of space, tests in drop towers with small free-fall test sections have shown significant effects on fluid behavior in two-phase flow in low-gravity environments of only $0.16-0.2 g$ and more limited experimental time durations. The 0.02-g acceleration in the KC-135, although not a perfect match to space conditions, is low enough to study low-gravity fluid behavior.

The COBRA/TRAC computer code was chosen to simulate the reduced-gravity data because it is an advanced best-estimate two-fluid computer code. It was developed to predict the thermal-hydraulic response of nuclear reactor primary coolant systems in severe accident transients. The code provides a two-fluid, three-field representation of two-phase flow, with a separate set of conservation equations for the continuous liquid, continuous vapor, and entrained droplet field. The equations are solved using a semi-implicit finite-difference technique on an Eulerian mesh. The code has extremely flexible noding capabilities for both the hydrodynamic field and the heat-transfer solution.

Closure of the conservation equations is achieved by using constitutive relations to define interfacial mass transfer, interfacial drag, wall drag, wall and interfacial heat transfer, entrainment and de-entrainment rates, and the

Table 1   Test matrix for boiling/condensing flow in reduced gravity

	Planned Test Conditions					As-Run Test Conditions			
Run no.	Flow rate (L/min)	Flow rate (kg/m²-s)	Power (W)	Exit quality	Void fraction	Parabola number	Flow rate (L/min)	Flow rate (kg/m²-s)	Average power (W)
1	0.91	307	3583	0.01	0.65	1	—	—	—
						2	0.8	270	3160
						3	0.8	270	3037
2	0.91	307	3934	0.02	0.75	4	0.7	237	3305
						5	0.7	237	3110
						6	0.6	203	—
3	0.91	307	4987	0.05	0.83	7	0.8	270	5163
						8	0.8	270	5160
						9	0.8	270	4671
4	0.30	101	1194	0.01	0.65	10	0.4	135	1049
						11	0.4	135	998
						12	0.3	101	1164

(Table 1 continued on next page.)

Table 1 (cont.)   Test matrix for boiling/condensing flow in reduced gravity

	Planned Test Conditions					As-Run Test Conditions			
Run no.	Flow rate (L/min)	Flow rate (kg/m²-s)	Power (W)	Exit quality	Void fraction	Parabola number	Flow rate (L/min)	Flow rate (kg/m²-s)	Average power (W)
5	0.30	101	2247	0.10	0.90	13	0.4	135	2725
						14	0.4	135	2783
						15	0.4	135	3037
6	0.45	152	3371	0.10	0.90	16	0.5	169	3780
						17	0.5	169	3323
						18	0.4	135	3228
7	0.45	152	2494	0.05	0.83	19	0.3	101	2203
						20	0.5	169	2510
						21	0.5	169	2630
8	0.45	152	1792	0.01	0.65	22	0.5	169	1902
						23	0.4	135	2064
						24	0.5	169	1919
9	0.45	152	5127	0.20	0.92		*		
10	1.20	405	4544	0.005	0.10		*		
11	1.20	405	4778	0.01	0.65		*		
12	1.20	405	5246	0.02	0.75		*		
13	0.30	101	1194	0.01	0.65		*		
14	0.45	152	5127	0.20	0.92		*		
15	1.20	405	4544	0.005	0.10		*		

*Experimental points not completed due to equipment failure.

Table 2   Test matrix for adiabatic two-phase flow in reduced gravity

Run no.	Planned Flow Rates				Parabola Number	As-Run Flow Rates			
	Water		Nitrogen			Water		Nitrogen	
	(L/min)	(kg/m²-s)	(L/min)	(kg/m²-s)		(L/min)	(kg/m²-s)	(L/min)	(kg/m²-s)
1	0.15	50	0.905	0.192	1	0.15*	50	—	—
					2	0.15*	50	—	—
					3	0.15*	50	—	—
2	0.75	248	0.905	0.192	4	0.6	198	0.9	0.2
					5	0.6	198	0.9	0.2
					6	0.5	165	0.9	0.2
3	0.15	50	3.016	0.641	7	0.15*	50	3.0	0.6
					8	0.15*	50	3.0	0.6
					9	0.15*	50	3.0	0.6
4	0.75	248	3.016	0.641	10	0.7	231	3.0	0.6
					11	0.7	231	3.0	0.6
					12		—	3.0	0.6
5	0.15	50	3.400	0.723	13	0.03*	10	0.9	0.2
					14	0.03*	10	0.9	0.2
					15	0.03*	10	0.9	0.2

(Table 2 continued on next page.)

Table 2 (cont.) Test matrix for adiabatic two-phase flow in reduced gravity

| | Planned Flow Rates | | | | | As-Run Flow Rates | | | | |
| | Water | | Nitrogen | | | | Water | | Nitrogen | |
Run no.	(L/min)	(kg/m²-s)	(L/min)	(kg/m²-s)	Parabola Number	(L/min)	(kg/m²-s)	(L/min)	(kg/m²-s)
6	0.03	10	0.905	0.192	16	0.15*	50	0.9	0.2
					17	0.15*	50	0.9	0.2
					18	0.15	50	0.9	0.2
7	0.03	10	3.016	0.641	19	0.7	231	6.1	1.3
					20	0.6	198	6.1	1.3
					21	0.5	165	6.1	1.3
8	0.03	10	3.400	0.192	22	0.024*	8	6.1	1.3
					23	0.024*	8	6.1	1.3
					24	0.024	8	6.1	1.3
9	0.75	248	6.075	1.29	25	0.6	198	6.1	1.3
					26	0.6	198	6.1	1.3
					27	0.6	198	6.1	1.3
10	0.30	99	3.016	0.641	28	0.16*	53	3.0	0.6
					29	0.16*	53	3.0	0.6
					30	0.16	53	3.0	0.6

*Below the range of the turbine flowmeter; reported from rotameter reading before parabola.

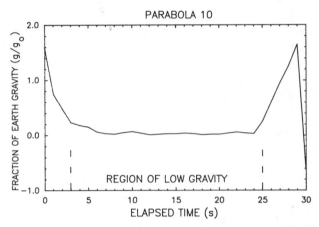

Fig. 3   KC-135 aircraft vertical acceleration during parabola 10.

thermodynamic properties of the fluid. COBRA/TRAC contains constitutive relations determined from physically based models of the relevant phenomena, but these are tailored to the conditions expected during postulated reactor transients. They are semiempirical, in that they are verified with experimental data, and are applicable only over a limited range of conditions.

The constitutive relations in the code are verified for reactor channel flow operational and transient conditions (see Ref. 5, Vol. 4). The conservation equations solved by COBRA/TRAC are applicable to the geometry and flowfield in the test section, and any solution difficulties can be attributed to the shortcomings of the constitutive relations when applied to reduced-gravity conditions. The purpose of modeling the experiment with COBRA/TRAC is to evaluate the applicability of the existing constitutive models to reduced-gravity conditions and to determine the areas where improved low-gravity constitutive models are needed.

Results from parabolas 2, 9, and 10 are reasonably representative of the range of conditions encountered in the boiling and condensing tests. Parabola 2 is at high flow rate and intermediate power; parabola 9 is at high flow rate and high power; and parabola 10 is at low flow rate and low power. These parabolas were modeled with COBRA/TRAC, and the code predictions are compared with the test data.

The COBRA/TRAC model of the test section is illustrated in Fig. 4. It consists of 31 fluid nodes, with 8 nodes in the boiler section, 8 nodes in the condenser, and 8 nodes in the condenser cooling jackets. Boundary conditions on the test section model were imposed by specifying the measured boiler inlet flow rate and the exit pressure at condenser measuring station C-P-10 (Fig. 2). (Post-test analysis indicated that the pressure recorded at the condenser exit, C-P-11, was the stagnation pressure rather than the static pressure; thus, it was deemed an inappropriate boundary condition for the simulation). The nodes modeling the condenser cooling jackets were

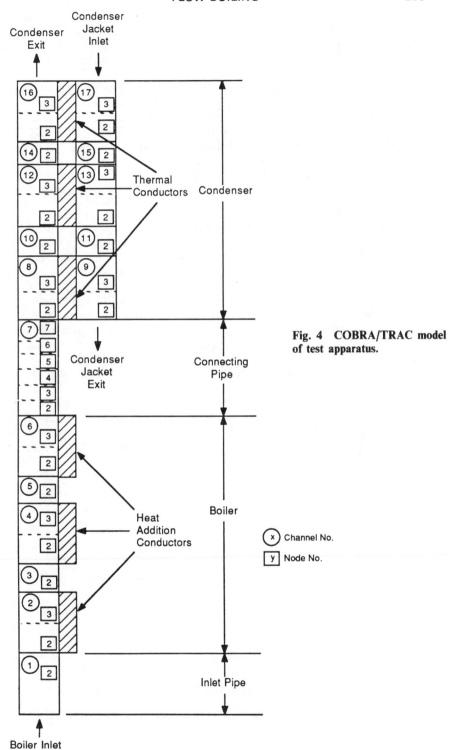

Fig. 4 COBRA/TRAC model of test apparatus.

given a constant inlet flow boundary condition of 1.14 l/min. The power input to the boiler was modeled with a heat flux boundary corresponding to test measurements of the power applied to the wires on the outer wall of the quartz tube.

Reduced gravity was simulated in the code by setting the gravity terms in the momentum equations to zero. No attempt was made to modify the value of g to follow the actual accelerometer readings for a given parabola. Constitutive relations that use the gravity acceleration constant g were not changed, nor was the existing vertical flow regime map. The COBRA/TRAC calculated results are compared to the time-dependent measured pressures and temperatures in the test section.

Parabola 2 Comparisons

Figure 5 shows the calculated temperatures in the boiler compared to the measured temperatures during the reduced-gravity interval. The calculated liquid temperature is plotted, since it is believed that the measured temperature is from a fully wetted thermocouple. A plot of the calculated saturation temperature at the boiler exit is included for comparison. Both the measured and calculated temperatures increase with axial length but remain constant with time at a particular measurement point. The exit temperature predicted by COBRA/TRAC is slightly higher than the measured value, but the calculations are generally in good agreement with the measured data.

Figure 6 shows the pressure drop across the boiler during the reduced-gravity interval. The data show variation with time, which is consistent with the flow pattern of intermittent surges of slugs of two-phase fluid observed in the high-speed video on similar test runs. These bubbly slugs appeared to bridge the boiler tube with a highly irregular and chaotic interface. The slugs were separated by intervals of relatively smooth

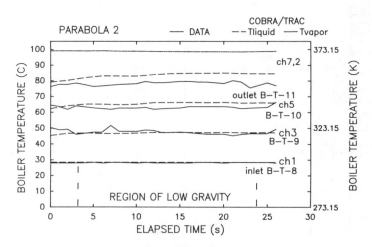

Fig. 5   Measured and calculated boiler temperatures for parabola 2.

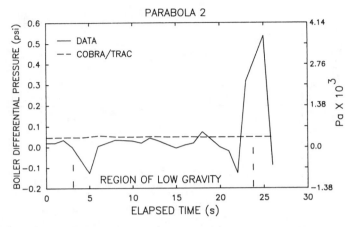

Fig. 6   Measured and calculated boiler pressure drop for parabola 2.

thin-film annular boiling flow. It is reasonable to assume that the pressure drop fluctuations correspond to the flow regime transitions associated with the formation and traverse of the boiler by the slugs. (The large measured variations at the beginning and end of the reduced-gravity interval are due to transitions into and out of the low-gravity region.)

The COBRA/TRAC calculation appears to be completely insensitive to the fluctuations in the flowfield, since the drag relations in the code do not correlate this behavior. The code predicts a bubbly flow regime with a smooth unvarying pressure drop that is in general slightly larger than what was actually observed in the boiler.

The calculated temperatures and pressures in the condenser during parabola 2 are compared to the measured values in Figs. 7 and 8,

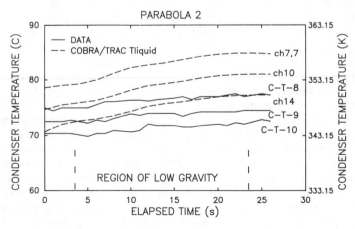

Fig. 7   Measured and calculated condenser temperatures for parabola 2.

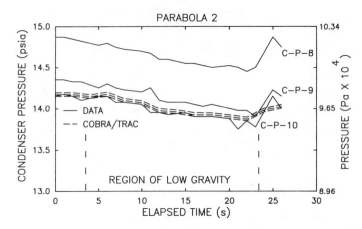

Fig. 8   Measured and calculated condenser pressures for parabola 2.

respectively. These results indicate significant differences in the condensation process. Neither the temperature nor pressure predictions matches the data well.

The COBRA/TRAC calculations predict a gradual increase in liquid temperature with time. This could be due to predicting a higher condensation rate than actually occurred, or to a lower wall heat-transfer rate. The former would cause the liquid temperature to increase due to the mass transport of condensed saturated vapor across the interface; the latter would result in increasing liquid temperature as insufficient heat was removed from the liquid through the condenser wall to the cooling jacket fluid. A combination of both factors could also produce this result.

The pressures predicted with COBRA/TRAC in the condenser are considerably lower than the corresponding measured values. These results indicate that the code does not predict a large enough pressure loss in the condenser. This is surprising, because the observed bubbly flow pattern in the condenser in reduced gravity has been predicted by COBRA/TRAC, and one would expect the drag forces to be similar.

Examining the measured pressure values in terms of pressure drop, it can be seen that the pressure loss in the condenser is approximately 5.17×10^3 Pa (0.75 psi), whereas the code predicts a pressure drop of 0.345×10^3 Pa (0.05 psi). The measured pressure drop seems unusually large for the flow conditions, since the equivalent single-phase pressure drop in the condenser for this flow rate is only 0.069×10^3 Pa (0.01 psi). This means that the two-phase friction pressure drop multiplier must be on the order of 75 to produce the observed pressures in the condenser. (The equivalent multiplier based on the pressure drop calculated in COBRA/TRAC is approximately 5.) This does not seem reasonable, and there is no particular phenomenon unique to 0-g conditions that would account for this extraordinarily high two-phase pressure drop.

The results shown here seem to indicate an error in the data either in the instrumentation or the raw data reduction algorithm. It is also possible that there is some factor at work that has not been taken into account in the code calculaton or the subsequent analysis. At this point one can only conclude that the repeatability of this phenomenon bears further careful study.

Parabola 9 Comparisons

Parabola 9 was at essentially the same flow rate as parabola 2, but at higher power, and consequently had a higher void fraction. The measured data and the COBRA/TRAC simulations for this test are presented in Figs. 9–12. The boiler temperatures calculated in COBRA/TRAC are compared to the measured values in Fig. 9. The code overpredicts the exit temperature, but the calculations match the trend of the data.

The boiler pressure drop is plotted in Fig. 10. The measurements show rapid fluctuations in the pressure. The COBRA/TRAC prediction shows less variability in the magnitude of the pressure drop. COBRA/TRAC predicts a bubbly flow regime in the boiler, except at the boiler exit, where annular flow is predicted. The high-speed video from this test shows a stable annular film boiling pattern near the boiler exit. At the higher void fraction in parabola 9, the constitutive relations for drag are predicting changes in the interfacial drag that are closer to the actual physical behavior observed. This implies that some work in flow regime mapping could greatly improve the capability for predicting pressure drop in two-phase flow in reduced-gravity near this range of flows and void fractions.

The condenser results are plotted in Figs. 11 and 12. COBRA/TRAC predicts considerably higher temperatures in the condenser than the measured data. The calculated exit temperature provides evidence of a local flow reversal. The increasing ramp in data temperature at all three

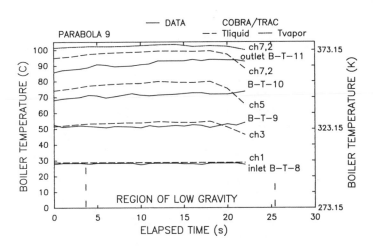

Fig. 9 Measured and calculated boiler temperatures for parabola 9.

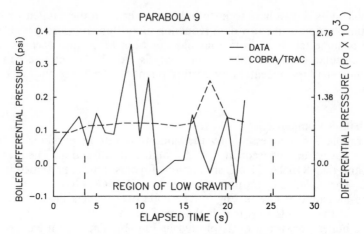

Fig. 10   Measured and calculated boiler pressure drop for parabola 9.

measuring stations indicates that flow in the condenser was not at thermal equilibrium in this test. The pressure data are essentially constant at each measuring station, indicating that momentum equilibrium is well established (Fig. 12).

COBRA/TRAC predicts annular flow at all condenser locations except channel 14 and calculates large fluctuations in pressure in the condenser during the reduced-gravity interval. These oscillations are probably due to inappropriate regime transitions being predicted, resulting in changes in predicted drag, based on the existing flow regime map in the code. The flow reversal at the condenser exit predicted by the code was not observed in the experiment and probably accounts for most of the discrepancy between the measured and calculated results.

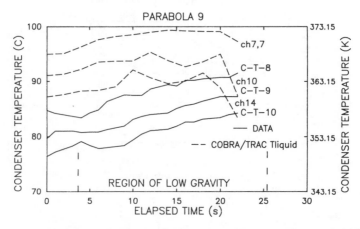

Fig. 11   Measured and calculated condenser temperatures for parabola 9.

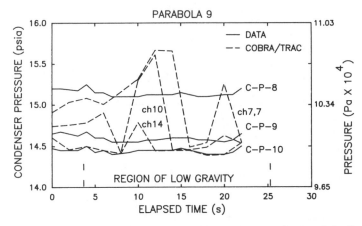

Fig. 12   **Measured and calculated condenser pressures for parabola 9.**

Parabola 10 Comparisons

Parabola 10 was at low flow rate and low power. The temperatures measured in the boiler, which are shown in Fig. 13, are clearly *not* the liquid phase temperature. The calculated boiler exit temperature matches the data very closely. The behavior shown at the intermediate stations (B-T-9 and B-T-10) is obviously due to repeated partial dryout and rewetting of the thermocouple. Both the high-speed video and the COBRA/TRAC prediction indicate a bubbly flow regime near the boiler exit.

At this low flow rate and void fraction, COBRA/TRAC predicts a steady pressure drop in the boiler, as shown in Fig. 14. Following the same pattern as in the other tests, the measurements show rapid changes in the

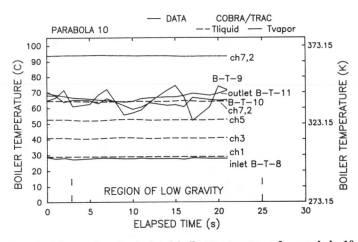

Fig. 13   **Measured and calculated boiler temperatures for parabola 10.**

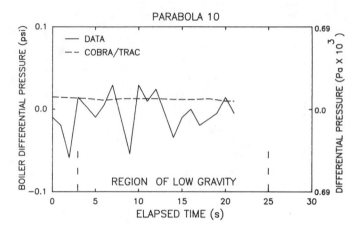

Fig. 14   Measured and calculated boiler pressure drop for parabola 10.

magnitude of the pressure drop. As with parabola 2, the predicted pressure drop is generally higher than the measured values.

In the condenser, the data for this test show that thermal equilibrium had been attained in the reduced-gravity interval (Fig. 15). The COBRA/TRAC predictions are slightly high at the condenser inlet. Since the calculated temperature was in better agreement with the measured value at the boiler exit, most of this difference is due to heat loss of 2–3°C in the line connecting the boiler to the condenser, which was not modeled in the code. The calculated temperature matches the measured temperature at station C-T-9, indicating that the code has calculated approximately the correct heat removal from the coolant in the first section of the condenser. But the

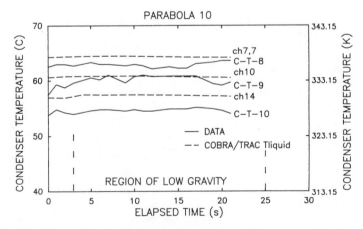

Fig. 15   Measured and calculated condenser temperatures for parabola 10.

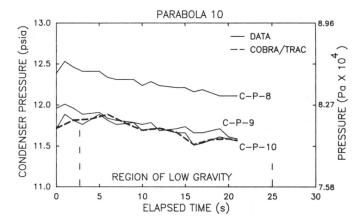

Fig. 16   Measured and calculated condenser pressures for parabola 10.

predicted temperature is high compared to measurements at the end of the second condenser section (C-T-10), indicating that not enough heat has been removed here. It is not clear whether this is due to actual change in the rate of heat removal from one section of the condenser to another, or to other effects.

Comparison of the calculated and measured pressures in the condenser, shown in Fig. 16, again shows a very small pressure drop in the bubbly flow regime predicted by COBRA/TRAC, while significant pressure drop is shown by the data. The same comments noted in the discussion of parabola 2 are relevant here. The cause of the large pressure drop in the condenser is not clear, and the phenomenon bears further investigation.

Flow Regime Mapping

The flowfield was recorded on a Kodak high-speed video motion-analyzer system and with NASA high-speed cameras in 11 of the 24 parabolas. Summaries of the visual observations are indicated on Table 3. These points have been plotted in Fig. 17 using the revised flow regime map developed by Dukler et al.[7] from reduced-gravity data obtained in the NASA Learjet and in drop tower tests. When plotted with superficial gas velocity vs superficial liquid velocity, the KC-135 data fall very nicely within the regime transition boundaries of Dukler and co-workers. The one exception is a test that appears to consist of bubbly flow, whereas the flow regime map predicts annular flow.

For very low gas velocity at both high and low liquid velocities, the observed flow regime was clearly a bubbly flow—small bubbles dispersed nearly uniformly in a liquid field. At higher gas velocities and somewhat lower liquid velocities, the flow consisted of well-defined slugs separated by regions of distinct annular flow. The slugs were in most cases very "frothy," containing many entrained bubbles, but the slugs contained enough liquid

Table 3 Flow regime comparisons

Parabola no.	Flow (L/min.)	Power (W)	BOILER		CONDENSER	
			COBRA/TRAC prediction	Experimentally observed near exit[a]	COBRA/TRAC prediction	Experimentally observed[b]
10	Low (0.4)	Low (1049)	One phase with bubbly near exit	Bubbly	Bubbly	—
19	Low (0.3)	Low (2203)	Half one phase, half bubbly	—	Annular	—
23	Low (0.4)	Low (1902)	—	Annular/slug/dryout	—	—
14	Low (0.4)	Intermediate (2783)	—	—	—	Annular
15	Low (0.4)	Intermediate (3037)	—	Annular/froth/dryout	—	—
6	Intermediate (0.6)	Intermediate (3200)	—	Annular/long slug	—	—
17	Intermediate (0.5)	Intermediate (3323)	—	Annular/froth/dryout	—	Annular

(Table 3 continued on next page.)

Table 3 (cont.) Flow regime comparisons

Parabola no.	Flow (L/min.)	Power (W)	BOILER COBRA/TRAC prediction	BOILER Experimentally observed near exit[a]	CONDENSER COBRA/TRAC prediction	CONDENSER Experimentally observed[b]
21	Intermediate (0.5)	Intermediate (2630)	—	Annular/slug/some dryout	—	—
2	High (0.8)	Intermediate (3160)	Bubbly	—	Bubbly	—
3	High (0.7)	Intermediate (3037)	—	—	—	Bubbly
5	High (0.7)	Intermediate (3110)	Bubbly with annular near exit	—	Bubbly at start and annular at end of parabola	—
7	High (0.8)	High (5163)	—	—	—	Bubbly
9	High (0.8)	High (4671)	Bubbly with annular near exit	Annular	Annular	—

Flow (kg/m²-s)
Low: < 150 (0.45 L/min.)
Intermediate: 150 < $\dot{m}$ < 215
High: > 215 (0.65 L/min.)

Power (W)
< 2300
2300 < $\dot{q}$ < 3500
> 3500

[a]Observed at point about 80% of total boiler length.
[b]Observed at center of condenser length.

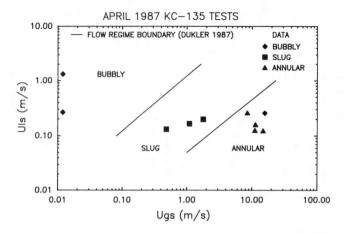

Fig. 17   KC-135 data points in Dukler's reduced-gravity flow regions map.

to clearly differentiate them from regions of annular flow with an exceptionally thick film, and very chaotic interfaces. At the highest gas velocities attained in these tests, the flow was indisputably annular, although on occasion interfacial waves were observed to momentarily bridge the tube entirely. The single data point labeled as "bubbly" that appears in the annular flow regime may be the result of misinterpretation of the visual records or errors in data readings. It is possible that this anomalous point was in annular flow, but with a thick film containing many small bubbles.

Table 3 summarizes the flow regime experimental observations and COBRA/TRAC predictions. The results of the simulations of these data with COBRA/TRAC show very clearly that the normal-gravity constitutive models in these codes, particularly those for interfacial drag, are not general enough to be applied indiscriminately to reduced-gravity two-phase flow. It appears that the flow boiling process itself is not very much different; hence, it is possible to correctly predict the heat-transfer and local temperatures if appropriate reduced-gravity flow regime maps are employed. However, the interfacial momentum exchange is not captured correctly, and new constitutive relations are needed for interfacial drag in addition to a corrected flow regime map. Differences are more clearly evident in condensing flow, since the data indicate that the actual pressure drop is much larger than those predicted with current constitutive relations. There is also some indication that condensation heat-transfer rates might be substantially different in reduced gravity, as compared to 1 g.

Needs for Two-Phase Thermal Hydraulics in Reduced Gravity

The data needs for two-phase flow in reduced gravity are twofold: the phase interaction for momentum and mass exchange and boiling heat transfer. Phase interaction can be characterized by defining flow regime maps and developing relations to describe the interfacial drag and heat transfer between the phases within each regime. Experimental measurements needed to develop these sorts of correlations include pressure drop, void fraction, phase velocities, and possibly turbulence intensities. Experimental results (including those discussed earlier), indicate that annular flow is the dominant flow pattern in reduced gravity, even for conditions that produce bubbly or slug flow in 1 g. However, more work is needed to characterize the flow regime over a wider range of flow rates and void fractions and to provide enough data to derive appropriate empirical relations for interfacial drag. Entrainment models to characterize the rate of removal of liquid drops from the continuous liquid phase will also be needed.

Development of heat-transfer correlations for boiling in reduced gravity will require data over a wide range of conditions and may prove more difficult to obtain than flow data, mainly for logistical reasons. The time required to achieve thermal equilibrium in terrestrial reduced-gravity facilities (such as drop towers and aircraft in parabolic flight) will be too short for some ranges of conditions that must be investigated, and power limitations of these facilities will impose further constraints on the range of data that can be investigated. In the long run, extended experiments in an orbital environment (such as onboard the Shuttle) will be required to develop a reliable wide-range set of correlations to characterize the full boiling curve.

For initial terrestrial investigation, however, it should be possible to accurately characterize the heat-transfer coefficients applicable to the nucleate boiling regime, which is the regime in which two-phase systems should be operating. The bubble nucleation at the heating surface is probably not significantly different in reduced gravity, since this takes place on the microscale, where the influence of body forces is likely to be minimal. But the mechanism of bubble departure from the surface, which is the main source of the high efficiency of heat transfer in nucleate boiling, is of necessity quite different. In 1 g the buoyancy force plays a major role in determining bubble behavior in flow boiling, whereas in 0 g the buoyancy force is almost completely absent, and surface tension forces are the dominating factor.

This raises several important questions about nucleate boiling in 0 g. How effectively will it transfer heat in reduced gravity? Previous work by PNL and other investigators has shown the prevalence of the annular flow regime, with the bubbles formed on the heated wall migrating rapidly through the liquid film on the wall to merge with the vapor core. The liquid film is unusually thick in reduced gravity, however, and the liquid temperature may be higher than it would be under similar conditions in 1 g. Also, the bubbles have been observed to get quite large before departing from the heated surface. This raises the question of how quickly the boiling process will encounter the critical heat flux (CHF) limit to nucleate boiling, since local vapor blanketing of the surface is one postulated mechanism of CHF for low-quality flow. The observed bubble behavior may indicate that CHF will occur at a lower heat flux in 0 g under some conditions. However, the thick liquid film holds out the possibility that, at high quality flows, where the usual mechanism of CHF is film dryout, the CHF point may occur at a higher heat flux.

Careful experimentation will permit investigation of nucleate boiling heat transfer over a range of wall superheats up to the point at which CHF occurs. The results of these tests will provide information needed to plan investigations of the CHF itself, in order to determine suitable design correlations for this important heat-transfer limit. The occurrence of CHF is a semistochastic transient event, and it may not be possible to control the flow boiling process precisely enough to evoke it reliably in the brief interval of reduced gravity in terrestrial facilities. Full investigations of CHF will have to be done in a longer-duration reduced-gravity environment, such as that available in the Space Shuttle.

Experiments in facilities such as the KC-135 can be performed for the range of the nucleate boiling region that can be covered within the power constraints of the facility, to obtain a limited amount of CHF data (probably in the form of test section burnout occurrences). Experience in 1 g has shown that nucleate boiling correlations generally extrapolate reasonably well over a large range of flow rates and pressures; hence, even correlations obtained for a relatively small data set may be useful for design purposes. The CHF data may be useful in defining conservative limit lines similar to those developed in the early design of nuclear reactor cores.

Behavior of Various Fluids in Reduced Gravity

Water is by most measures the best working fluid for thermal-hydraulic systems. It is a benign fluid, noncorrosive when free of impurities, and its thermophysical properties are very well known. But in some applications, other working fluids may have properties that make them more suitable than water. For example, liquid alkali metals have a much larger heat-carrying capacity than water; thus, a system with such a coolant could be much smaller than one designed for water, yet still produce the same power. Organic coolants sometimes have advantages over water in low-pressure and temperature-operating regions, particularly in refrigeration systems. The behavior of fluids other than water in reduced gravity is a significant concern for design of thermal-hydraulic systems for use in space.

The same concerns discussed throughout this text for water apply to any other coolant that will be operated in a two-phase loop in reduced gravity. The experience in 1 g has been that correlations and empirical models developed for water do not generally adequately describe the behavior of alternative fluids. One notable exception is the behavior of the Freon refrigerants, which do behave much the same as water, when proper account is taken of scaling factors. Ammonia is another coolant with properties similar to water that is being considered for space applications where escaping vapors will not pose an unacceptable environmental hazard. But, in general, it is not reasonable to expect reduced-gravity correlations developed for water to apply indiscriminately to the behavior of other fluids.

This will be particularly important for fluids with surface tension significantly different from that of water, since this property has a very large effect on fluid behavior in reduced gravity.

Reduced-gravity flow and heat-transfer correlations and empirical models developed for water can be applied to other fluids only when adequate experimental data are available for verifying them with the alternative fluid. It will be necessary to first verify that the correlation or model has the appropriate terms to describe the behavior of the new fluid. Then enough data must be obtained, over an appropriate range of conditions, to determine values for the empirical coefficients applicable with the particular fluid. The amount of data actually needed to obtain adequate correlations for fluids other than water may not be as extensive as that envisioned for a complete database for water, but it may be more difficult to obtain in practice.

Safety concerns may limit the testing permitted with some fluids in some reduced-gravity facilities. In aircraft test flights at high altitudes and in the Space Shuttle, there is an understandable reluctance to deal with fluids that can explode, release poisonous vapors, or otherwise pose an active danger to the craft or operating personnel. Power limitations of the test facility may also limit the range of data that can be obtained in reduced gravity, resulting in unacceptable uncertainty as to the applicability of correlations and models in the actual operating range of space thermal-hydraulic systems. Tests in an environment that can currently be provided only by the Space Shuttle will be necessary for any fluid that will be used in large power systems in space. This is true even for water, and it is particularly true for liquid metals, with the added concern that the material must be handled safely.

Differences in behavior among various working fluids are generally due to differences in the thermodynamic and transport properties of the fluids. Liquid metals have very high thermal conductivities and as a result can transfer heat at a much higher rate than a fluid that relies mainly on convection alone to remove heat from a surface. Surface tension governs to a large degree the process of bubble formation in flow boiling, determining the size of bubbles, and the ease with which they can form, thus affecting the rate at which nucleate boiling heat transfer can remove heat from the surface. A fluid that forms many small bubbles will transfer heat at a different rate than one that more easily forms large bubbles.

Cryogenic fluids are not in precisely the same category as thermal-hydraulic system coolants, but in some applications in reduced gravity they can be expected to undergo two-phase flow, e.g., in tank transfer and chilldown operations and in vaporization for introduction into a rocket engine combustion chamber. Design and analysis tools such as two-fluid computer codes to model these conditions need constitutive models capable of describing the phase interactions of cryogenic liquid and vapor and the boiling process when the cryogenic fluid comes in contact with a relatively "hot" wall. Much of these data can be obtained in terrestrial facilities, but some specialized applications may also require testing in the Shuttle as well.

References

[1]Antoniak, S. I., "Two-Phase Alkali-Metal Experiments in Reduced Gravity," Pacific Northwest Lab., Richland, WA, PNL-5906, June 1986.

[2]Krotiuk, W. J. and Cuta, J. M., "Low Gravity Fluid-Thermal Experiments," presented at the Third Pathways to Space Experimental Workshop, Orlando, FL, June 3, 1987.

[3]Kachnik, L. J., "Design, Construction, and Testing of a Boiling and Convective Condensation Experiment for Use in a Microgravity Environment," Ph.D. Thesis submitted to Texas A&M University, 1987.

[4]Kachnik, L. J., Lee, D., Best, F., and Faget, N., "A Microgravity Boiling and Convective Condensation Experiment," American Society of Mechanical Engineers, TP 87-WA/HT-12. Dec. 1987.

[5]Thurgood, M. J., George, T. L., Guidotti, T. E., Kelly, J. M., Cuta, J. M., and Kuontz, A. S., "COBRA/TRAC—A Thermal-Hydraulics Code for Transient Analysis of Nuclear Reactor Vessels and Primary Coolant Systems," Pacific Northwest Lab., Richland, WA, NUREG/CR-3046, PNL-4385, Vols. 1–5, March 1983.

[6]Best, F. and Faget, N., "Test Equipment Package for Microgravity Two-Phase Flow Project," NASA Johnson Space Center, Houston, TX, 1986.

[7]Dukler, A. E., Fabre, J. A., et al., "Gas Liquid Flow at Microgravity Conditions: Flow Patterns and Their Transactions," *International Symposium on Thermal Problems in Space Based Systems*, American Society of Mechanical Engineers, HTD-Vol. 83, New York, Dec. 1987.

Computation of Space Shuttle High-Pressure Cryogenic Turbopump Ball Bearing Two-Phase Coolant Flow

Yen-Sen Chen*
SECA, Inc., Huntsville, Alabama

Introduction

Computational investigation of the phase change and heat-transfer characteristics of the liquid oxygen coolant flow inside the pump-end ball bearing assembly of the high-pressure oxygen turbopump of the Space Shuttle main engine is presented. Computational fluid dynamics methodology is employed to solve the three-dimensional Reynolds-averaged Navier-Stokes equations in describing the fluid flow motion. A rotational coordinate mesh system with rotational forces implemented in the momentum equations is utilized in this study. A pressure-based flow solver with a pressure multicorrector time-marching scheme is employed for velocity-pressure coupling. A two-equation turbulence closure model is also used for turbulent flow computations. For simulating the phase change mechanism due to frictional heat flux near the Hertzian contact ellipse, a homogeneous two-phase model with bubble growth rate formulation is employed in conjunction with the table of oxygen property provided by the National Bureau of Standards.

Under various operating conditions (minimum, rated, and full power levels), the ball bearing assembly of the turbopump shaft of the Space Shuttle main engine (SSME) is always experiencing strong loadings (static and dynamic loads).[1] For instance, the pump-end ball bearing assembly (number 3 and number 4 bearings) of the high-pressure oxygen turbopump

*Research Scientist. Formerly Visiting Scientist, USRA/Fluid Dynamics Branch, NASA Marshall Space Flight Center.

275

(HPOTP) is constantly subject to loadings of around 600–1,800 lb for the number 4 and number 3 bearings (for full power level with shaft speed of 29,500 rpm) in addition to the 1,000-lb preloads. Because these bearings are lubricated with the cryogenic fluid liquid oxygen (LOX), there is essentially no damping effect provided by the coolant. Damping effect of the coolant is very important in reducing the dynamic loadings acting on the bearing. For the present coolant, large amounts of heat (due to large dynamic loadings) are generated in small areas of the Hertzian contact ellipse (i.e., the elliptic contact surface between the ball and the races) through contact friction between the ball and the races. With the addition of viscous heat generation, significant phase change or boiling zones of the coolant flow can be produced near the contact areas. At the entrance of the bearing coolant flow passage (the number 4 bearing end), the static pressure and temperature are approximately 335 psia and 203°R, respectively. This indicates a 17-Btu/lbm margin for the phase change process to occur. For a friction factor of 0.2 between the ball and the races, the frictional heat-generation rates for the number 3 and number 4 bearings (each bearing consists of 13 0.5-in.-diam balls circumferentially) are 10.32 and 9.44 Btu/s, respectively, for the full power level as provided by Rocketdyne's analysis.[2] It is the main objective of this study to investigate the possibility and the extent of phase change of LOX flow near the contact areas due to the frictional and viscous heat fluxes for the given inlet LOX conditions.

Effects of phase change of LOX may have significant impact on the heat-transfer characteristics between the ball bearing and the coolant. Local boiling zone tends to reduce the heat-transfer rate near the solid wall boundaries and hence reduces the overall cooling effect. To a large extent, the net effect may cause the ball bearing to go into a thermal runaway mode and to jeopardize the entire turbopump.

To model this problem, a three-dimensional Navier-Stokes flow solver[3-5] with a homogeneous two-phase model[6,7] is employed to describe the fluid motion inside the coolant passage. In the formal formulations of two-phase bubbly flow momentum equations with continuum mechanics analogy,[8] the interphase drag force due to velocity difference and the interphase stress terms due to bubble deformation may have significant effects on the bubble trajectory inside the flowfield. However, for simplicity, these effects are ignored in this study due to the complexity of the present three-dimensional flow problem. It is therefore assumed that the two phases share the same velocity field and there can only be one spherical bubble existing inside any control volume. The bubble growth rate model of Rohatgi and Reshotko[6] and Simpson and Silver[7] is then employed to control the evolution of the bubble inside the control volume based on the difference between the bubble pressure and the vapor pressure. The oxygen properties (e.g., temperature, density, thermal conductivity, viscosity, Prandtl number, etc.) inside the flowfield are obtained through a table provided by the National Bureau of Standard (NBS)[9] using the local fluid enthalpy and pressure as entries. This thermophysical properties table covers a range in temperature from freezing line to 600°R and a range in pressure from 1 psia to 5000 psia. The present application falls well inside the defined ranges. Figure 1 illustrates a sketch of the LOX phase diagram obtained from the NBS data.

Governing Equations

The basic equations employed to describe the flow motion of general Newtonian fluid follow certain conservation laws: conservation of mass, conservation of momentum, and conservation of energy. A transport equation written in generalized coordinates system is given to state these conservation laws:

$$(1/J)(\partial \rho q/\partial t) = \partial[-\rho U_i q + \mu_{\text{eff}} G_{ij}(\partial q/\partial \xi_j)]/\partial \xi_i + S_q \qquad (1)$$

where $q = 1$, u, v, w, h, k, ε, and m for the continuity, momentum, energy, k-ϵ turbulence model, and discrete-phase mass fraction transport equations, respectively; and S_q respresents source terms of the transport equations. These source terms are given as:

$$S_q = (1/J)\begin{Bmatrix} 0 \\ -p_x + \nabla[\mu_{\text{eff}}(u_j)_x] - \tfrac{2}{3}(\mu_{\text{eff}}\nabla u_j)_x \\ -p_y + \nabla[\mu_{\text{eff}}(u_j)_y] - \tfrac{2}{3}(\mu_{\text{eff}}\nabla u_j)_y \\ -p_z + \nabla[\mu_{\text{eff}}(u_j)_z] - \tfrac{2}{3}(\mu_{\text{eff}}\nabla u_j)_z \\ Dp/Dt + \Phi \\ \rho(P_r - \epsilon) \\ \rho(\epsilon/k)(C_1 P_r - C_2\epsilon) \\ \rho W_t \end{Bmatrix}$$

where p, Φ, P_r, and W_t stand for static pressure, energy dissipation function, turbulent energy production rate, and mass fraction generation due to bubble growth rate, respectively; C_1 and C_2 are turbulence modeling constants; and J, U_i, and G_{ij} represent the Jacobian of coordinate transformation, contravariant velocities, and diffusion metrics, respectively. The last three symbols are written as

$$J = \partial(\xi,\eta,\zeta)/\partial(x,y,z)$$
$$U_i = (u_j/J)(\partial \xi_i/\partial x_j)$$
$$G_{ij} = (\partial \xi_i/\partial x_k)(\partial \xi_j/\partial x_k)/J$$

where x_i and ξ_i are coordinate systems for the physical and transformed domains, respectively; $\mu_{\text{eff}} = (\mu + \mu_t)/\sigma_q$ is the effective viscosity when the turbulence eddy viscosity concept is employed to model the turbulent flows; $\mu_t = \rho C_\mu k^2/\epsilon$ is the turbulence eddy viscosity; and C_μ and σ_q denote

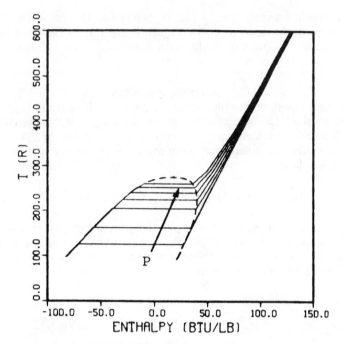

Fig. 1 Phase diagram of oxygen.

turbulence modeling constants. In this study, turbulence modeling constants given in Ref. 10 are used (i.e., $C_\mu = 0.09$, $C_1 = 1.15 + 0.25 P_r / \epsilon$, $C_2 = 1.90$, $\sigma_k = 0.80$, and $\sigma_\epsilon = 1.05$).

Bubble Growth Model

A detailed derivation of the present bubble growth rate two-phase model was given by Rohatgi and Reshotko.[6] This model is summarized in the following paragraphs. The basic equations used in deriving this model include the Clausius-Clapeyron equation and the simplified one-dimensional momentum and energy equations. The final expression can be written as

$$dm/dp = v_{fg}/h_{fg}(m - CT/h_{fg}) \qquad (2)$$

where m, p, v, h, C, and T represent the mass fraction (or quality) of the bubbly phase, pressure, specific volume, enthalpy, specific heat, and fluid temperature, respectively; the subscript fg denotes differences due to phase change.

Then a model of rate of production of heterogeneous nuclei per unit volume proposed by Simpson and Silver[7] was employed:

$$dn/dt = N_n(2\sigma/M\pi)^{\frac{1}{2}} \exp[-(16\pi\sigma^3\delta/3\kappa T)/(p_g - p_f)^2] \qquad (3)$$

where

N_n = number of heterogeneous nucleation sites per unit volume
σ = surface tension of oxygen
M = mass of a molecule
κ = Boltzmann's constant
δ = function of bubble contact angle (assumed to be unity in this study)
p_g = gas pressure at fluid temperature T
p_f = fluid pressure

With Eqs. (2) and (3), an expression for the bubble growth rate can be written as

$$\mathrm{d}r/\mathrm{d}t = [K_f R^2 T_f^2 / h_{fg}^2 p_g^2 (3Kt/\pi)^{\frac{1}{2}}](p_g^2/p_{gr})\ell n(p_g/p_{gr})$$

where

K_f = thermal conductivity of fluid
K = thermal diffusivity of fluid
p_{gr} = pressure inside a bubble of radius r
R = gas constant

Finally, the source term for the mass fraction equation can be written as

$$\rho W_t = \rho_g (1 - \alpha)(4\pi/3)(U/\Delta s)(\mathrm{d}n/\mathrm{d}t)(\mathrm{d}r/\mathrm{d}t)^3$$

where the volume fraction α can be calculated from the following homogeneous relation:

$$\alpha = m(\rho_f/\rho_g)/[1 + m(\rho_f/\rho_g - 1)]$$

Numerical Scheme

To solve the system of nonlinear coupled partial differential equations [Eq. (1)] finite-difference approximations are used to establish a system of linearized algebraic equations. A relaxation solution procedure is employed in the present study to couple the governing equations. For convenience, Eq. (1) is rewritten as

$$(1/J)(\partial \rho q/\partial t) = -\partial F_i/\partial \xi_1 + S_q = R_q \tag{4}$$

where F represents convection and diffusion fluxes. First, Eq. (4) is discretized in time with a time-centered (Crank-Nicholson) scheme:

$$(1/J\Delta t)[(\rho q)^{n+1} - (\rho q)^n] = (R_q^{n+1} + R_q^n)/2$$

where the superscript n denotes the current time level. If a subiteration procedure within a time step is applied, the following linearization can be incorporated:

$$(\rho q)^{n+1} = (\rho q)^k + \rho^n \Delta q^k$$

$$R_q^{n+1} = (\partial R_q / \partial q)^k \Delta q + R_q^k$$

where the superscript k denotes the kth subiteration. With the preceding approximations, the final form of the time-marching scheme can be written as

$$[(\rho^n / J \Delta t) - 1/2(\partial R_q / \partial q)^k] \Delta q^k = -(1/J \Delta t)[(\rho q)^k - (\rho q)^n] + (R_q^k + R_q^n)/2$$

The solution at time level $n + 1$ is then updated by

$$q^{n+1} = q^{k+1} = q^k + \Delta q^k$$

The expression $k = 1$ (i.e., a noniterative time-marching procedure) is selected in the present investigation. As reported in Refs. 4 and 5, a multicorrector solution method can provide time-accurate solutions for transient flow problems.

For spatial discretizations, an upwind scheme is employed based on a second-order central plus artificial dissipation scheme. For simplicity, let us consider fluxes in the ξ direction only. That is,

$$\partial F / \partial \xi \cong (F_{i+1} - F_{i-1})/2 - (d_{i+\frac{1}{2}} - d_{i+\frac{1}{2}})$$

where d denotes a dissipation term. The dissipation terms are constructed such that a second- or third-order upwind scheme is activated in smooth regions and a first-order upwind scheme is recovered near shock waves. Detailed formulations of the dissipation terms are given in Ref. 5.

A pressure-based multicorrector solution procedure is also employed in the present flow solver. A simplified momentum equation is combined with the continuity equation to form a pressure correction equation. This pressure correction equation reveals elliptic features for low-speed flows and continuously becomes hyperbolic as flow speed increases. The velocity and pressure fields are corrected based on the solution of the pressure correction equation. The density field is then updated by the oxygen property table. To ensure that the updated velocity, density, and pressure fields satisfy the continuity equation, the preceding pressure correction solution procedure is repeated several times (usually 4 times are sufficient) before marching to the next time step. This represents a multicorrector solution procedure. Successful applications of this method to transient two- and three-dimensional flow problems at low and high Reynolds numbers have been reported in Refs. 4 and 5.

To solve the system of linear algebraic equations, an alternating-direction line-relaxation matrix solver is incorporated in the present study.

Boundary Conditions

Various types of boundary conditions can be incorporated in the present computer code by simply changing the input data for boundary control parameters. For the present application, the entire flowfield is subsonic. At the flow inlet, only the pressure field is extrapolated. Other inlet flow conditions were obtained from the results of a three-dimensional inlet-jet-plenum numerical computation.[11] A 30-jet configuration was considered in that study, from which an averaged flowfield at the jet-plenum outlet was imposed for the present inlet conditions. Along the exit plane, which is downstream of the number 3 bearing, a mass conservation condition is imposed. Pressure distributions at the exit plane are obtained by linear extrapolation from the interior pressure solutions. For the circumferential boundaries (considering one-thirteenth sector of the entire bearing as the computational domain), a cyclic boundary condition is employed. It is also assumed in this study that the tandem bearings are operating at a condition in which each ball is lined up perfectly. This further simplifies the problem. Other configurations can be considered in future study. For the solid wall boundaries, conditions of no-slip for velocity and zero normal gradient for pressure are imposed. For turbulent wall boundary conditions, standard wall function approach[12] is employed for computational efficiency of the 3D flow problem.

Results and Discussion

The computational domain involved in the present test case consists of a coolant flow passage for the fluid part and the balls, inner and outer races, and the cage of the balls for the solid part. Figure 2 depicts the mesh system (with grid size of $71 \times 33 \times 31$) employed in the present investigation. The coolant flow enters the bearing from the left of the domain (i.e., the number 4 bearing end) and exists from the right (i.e., the number 3 bearing end). Frictional heat sources of ball/inner race, ball/outer race, and ball/cage are added near the locations with an x sign (see Fig. 2). The volume used to distribute frictional heat source to one contact point between the ball and the race is estimated to be 2.53×10^{-5} in.^{3} based on the area of contact ellipse and the assumed clearance of 2% ball diameter. This volume is assigned on the surfaces of the ball and the races. It is assumed that the flow can pass through the gap of the contact points to transport heat to other flow regions. For solution stability, the large rotational forces due to the high rotational speed of the system are included in the momentum equations with a gradual increase in magnitude as a function of time before reaching the final value. The outermost boundaries are assumed to be isothermal at the LOX inlet temperature. It took 1500 time steps (with nondimensional time step sizes of 0.001 for the fluid part and 0.1 for the solid parts) to obtain a converged solution. This corresponds to about 8 h (or 2.65×10^{-4} s/step/grid) of CPU time on the Cray-XMP supercomputer. It is learned that about 10% of the CPU time has been spent on the table look-up for the LOX properties.

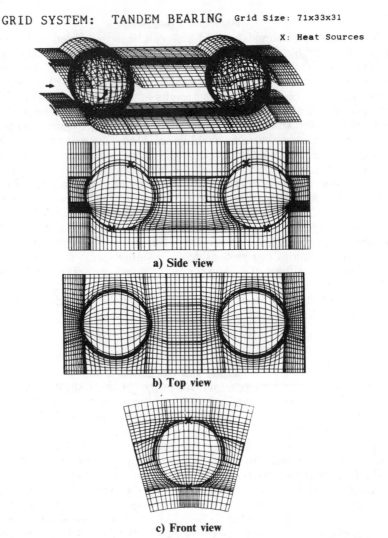

Fig. 2   Mesh system for the number 3 and number 4 (upstream) bearings.

Figure 3 shows the flowfield solution inside the flow passage of the bearing. It can be seen that the circumferential flow is dominating and the flow patterns near the ball/race contact points are quite complex. The flow speed is increased going through the narrow gap partially due to area reduction and partially due to phase change effect. Large viscous heat generation can be expected around contact points due to large velocity gradients. The flow is separated downstream of the expansion shoulder of the number 4 bearing. This is caused by the local geometrical feature and the adverse pressure gradients. This may reduce heat transfer along the inner-race surface.

The predicted temperature distributions for the entire flow/solid domain is given in Fig. 4. The isotherms show an increment in temperature of 50°R. Hot spots on the surfaces of the ball, the inner race, the outer race, and the cage are also given in Fig. 4. Temperatures of these hot spots are summarized in Table 1.

Except for the hot spots on the ball surface, the temperature inside the ball is quite uniform due to high rotational speed. The predicted average ball temperatures are 486 and 505°R for the number 4 and number 3 bearings, respectively. For the fluid part the highest temperature is 409.4°R, occurring near the ball/inner-race contact point of the number 3 bearing. Some latent heat has been absorbed for phase change. This and the temperature gradients required for heat transfer make the fluid temperature much lower than that of the nearby solid part.

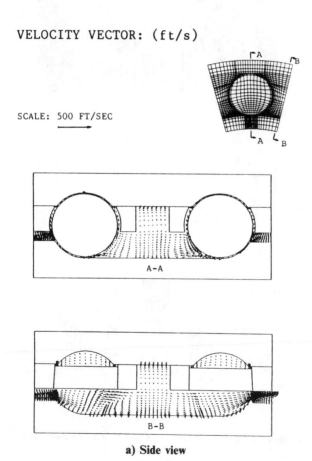

a) Side view

Fig. 3   Velocity vectors inside the bearing coolant flow passage.

(Figure continued on next page.)

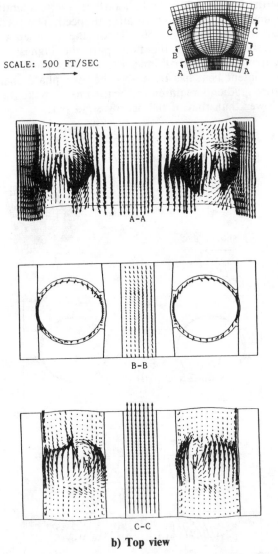

SCALE: 500 FT/SEC

A-A

B-B

C-C

b) Top view

Fig. 3 (cont.)   Velocity vectors inside the bearing coolant flow passage.

(Figure continued on next page.)

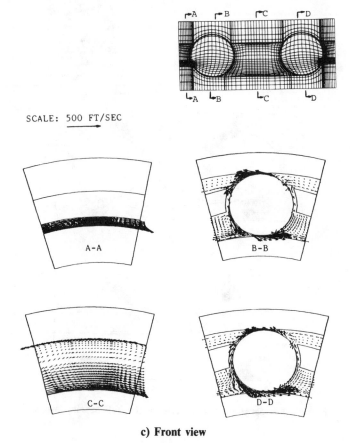

SCALE: 500 FT/SEC

c) Front view

Fig. 3 (cont.)   Velocity vectors inside the bearing coolant flow passage.

Table 1   Predicted hot spots (°R)

	No. 4	No. 3
Ball/inner race	1624	1809
Ball/outer race	811	901
Ball/cage	709	780

TEMPERATURE FIELD: (°R)

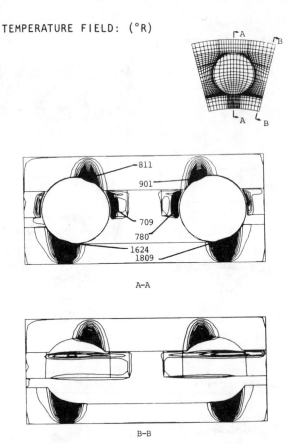

Fig. 4   **Temperature distributions of the ball bearing assembly.**

Summary

A homogeneous two-phase fluid flow model has been implemented in a three-dimensional Navier-Stokes solver. Application of this method to the pump-end bearing coolant flow analysis of the HPOTP of the SSME has been presented. Results of the present analysis have indicated large boiling zones and hot spots near the ball/race contact points. The present method may provide useful information for bearing coolant flow design and analysis. Detailed flowfield surrounding the ball bearing and heat-transfer characteristics on the ball surface can be studied with the present approach. This will provide a better understanding of a ball bearing design. Consequently, a design change may be suggested.

Acknowledgments

The author would like to acknowledge Dr. Nicholas Costes, the technical monitor, for his support and helpful suggestions. This work was supported by NASA Contract NAS8-35918 with the Universities Space Research Association. Additional support was provided by SECA, Inc.

References

[1]Childs, D. W. and Moyer, D. S., "Vibration Characteristics of the HPOTP (High-Pressure Oxygen Turbopump) of the SSME (Space Shuttle Main Engine)," *Journal of Engineering for Gas Turbines and Power*, Vol. 107, Jan. 1985, pp. 152–159.

[2]Rankel, R. E., "HPOTP 'Turbine-End' Bearings Thermal Analysis Parameters," Rocketdyne Rept., Rocketdyne, IL, Attachment 1 (SSME-87-461) and Attachment 2 (SSME-86-2286), March 1987.

[3]Chen, Y. S., "Viscous Flow Computations Using a Second-Order Upwind Differencing Scheme," AIAA Paper 88-0417, Jan. 1988.

[4]Chen, Y. S., "3-D Stator-Rotor Interaction of the SSME," AIAA Paper 88-3095, July 1988.

[5]Chen, Y. S., "Compressible and Incompressible Flow Computations with a Pressure Based Method," AIAA Paper 89-0286, Jan. 1989.

[6]Rohatgi, U. S. and Reshotko, E., "Non-Equilibrium One-Dimensional Two-Phase Flow in Variable Area Channels," *Proceedings of the Institute of Mechanical Engineers*, Institute of Mechanical Engineers, 1976.

[7]Simpson, H. C. and Silver, R. S., "Theory of One-Dimensional, Two-phase Homogeneous Non-equilibrium Flow," *Proceedings of the Institute of Mechanical Enginneers*, Institute of Mechanical Engineers, 1962, pp. 45–56.

[8]Drew, D. A., "Mathematical Modeling of Two-Phase Flow," *Annual Review of Fluid Mechanics*, Vol. 15, 1983, pp. 261–291.

[9]McCarty, R. D. and Weber, L. A., "Thermophysical Properties of Oxygen from the Freezing Liquid Line to 600 R for Pressures to 5000 psia," National Bureau of Standards, NBS-TN-384, July 1971.

[10]Kim, S. W. and Chen, Y. S., "A Finite Element Computation of Turbulent Boundary Layer Flows With an Algebraic Stress Turbulence Model," *Computer Methods in Applied Mechanics and Engineering*, Vol. 66, Jan. 1988.

[11]Chen, Y. S., "Homogeneous Two-Phase Flow Calculation Inside the HPOTP Ball Bearing Assembly," presented at the Sixth SSME Working Group Meeting, NASA Marshall Space Flight Center, April 26–28, 1988.

[12]Launder, B. E. and Spalding, D. B., "The Numerical Computation of Turbulent Flows," *Computer Methods in Applied Mechanics and Engineering*, Vol. 3, 1974, pp. 269–289.

Chapter 7. Reduced-Gravity Condensation

Reduced-Gravity Condensation

Frederick R. Best*

Texas A&M University, College Station, Texas

Nomenclature

c_f = friction factor constant
c_p = specific heat
d = tube inner diameter
F = Henstock and Hanratty's dimensionless interfacial factor[12]
G = mass velocity
$f_{1,2,3,4,5}$ = factors from Pohner's condensation model[13]
g = acceleration
h_{fg} = latent heat of vaporization
h'_{fg} = corrected latent heat of vaporization
i = interface subscript
k = conductivity
l = liquid
L = length
m = flow rate
R = tube outer radius
Re = Reynolds number
s = slip ratio; vapor-to-liquid velocity ratio
T_s = vapor saturation temperature
T_w = wall temperature
t = time
u_v = vapor velocity
u_{v0} = entrance vapor velocity
u_{l_i} = liquid film velocity at the interface
z = flow direction

*Associate Professor, Nuclear Engineering Department.

δ = film thickness
γ = Henstock and Hanratty's dimensionless film parameter
μ = viscosity
ρ_l = liquid density
ρ_v = vapor density
τ = shear

Introduction

The National Research Council's Committee on Industrial Applications of the Microgravity Environment[1] reported that, although near-term space manufacturing is unlikely, "commercialization of the microgravity environment will depend upon the success of basic research projects performed in space." Among space research platforms, the Space Shuttle can supply a nominal 3.5 kWe from fuel cells, and the Space Station is projected to require about 75 kWe in phase 1. As space missions turn from exploration and discovery to commercial ventures, power requirements greater than 400 kWe are anticipated.[2] The Arthur D. Little Company[3] projects that photovoltaic systems will continue to play an important part in space power systems but that, for systems requiring 100 kWe and higher, dynamic conversion systems will become important. In addition, spacecraft have thermal energy management requirements as well as electrical power requirements.

Motivation for Interest in Microgravity Condensation

Thermal energy is produced within spacecraft as electrical energy degrades to waste heat, as chemical/mechanical processes produce waste heat, and from crew environmental effects. Furthermore, solar insolation may necessitate thermal energy management, and spacecraft process thermal loads for manufacturing will one day be significant.

Dynamic Power Conversion and Thermal Energy Management

The need for increased electrical power and thermal energy management capability has led spacecraft designers to consider more mass-efficient energy management systems than the conduction, thermal radiation, and single-phase convective systems presently in use. Two-phase thermal energy transport devices called heat pipes have been used in spacecraft for years. However, pumped two-phase systems, having vapor and liquid phases as the principal energy transport media, are leading candidates for future energy management systems. Two-phase thermal systems intrinsically depend on boiling/evaporation and condensation processes for operation. Chapters 4 and 6 discuss boiling processes. This chapter will deal with condensation in a microgravity environment. The objectives of this chapter are 1) to present the theoretical background for condensation phenomenology, 2) to summarize the available condensation experimental data, 3) to compare theory with experiments, 4) to describe planned research projects, and 5) to make recommendations for further work.

Theoretical Background

This section will present the theoretical development for various condensation processes under microgravity conditions. Several preliminary comments regarding microgravity must first be made. The term microgravity has come to have several meanings. Literally, microgravity could mean $10^{-6} \times$ standard Earth gravity, it could mean a reduced-gravity acceleration, or even a true 0 *g*. This proliferation of meanings has evolved because of the differing viewpoints of space researchers. Crystal growth and electrophoresis are types of experiments for which true *μ-g* acceleration conditions are required. However, reduced-gravity acceleration conditions are of interest for systems under continuous boost or on celestial bodies with gravity less than that of the Earth's, e.g., the moon (1/6 *g*).

Researchers in heat transfer and fluid flow have developed some models that use the local acceleration explicitly as a variable and other models that are based on a complete absence of buoyancy forces, true 0 *g*. Throughout this chapter microgravity will be taken to mean some explicit reduced-gravity value, not exactly 10^{-6} standard Earth gravity.

Figure 1 shows a general condensation process in which high-energy vapor phase molecules interact with the interface of a bulk liquid phase, either joining it or rebounding from it. High-energy liquid molecules may also escape into the vapor region. The fundamental condensation process is not thought to depend on the acceleration field. This will be assumed to be correct even though no experiments have directly investigated the validity of this assumption.

The terms "condensation" or "condensation processes" will be used to describe the macroscopic approach taken in this chapter in which a bulk vapor phase interacts with a bulk liquid phase, with the condensation rate being influenced by the relative geometry of the two phases.

Phenomenology

There are two fundamentally different condensation processes: homogeneous condensation and inhomogeneous condensation. Homogeneous

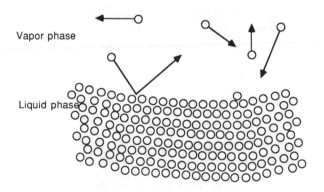

Fig. 1 Fundamental condensation schematic.

condensation occurs in the bulk phase of a vapor when molecules of sufficiently low energy coalesce to form microscopic liquid phase particles. Although this process might be of interest for certain spacecraft situations, e.g., vapor leaking into the space vacuum or the air-handling system of environmental life support, homogeneous nucleation does not play a part in condensation occurring in process equipment and therefore will not be discussed further.

Inhomogeneous condensation occurs when bulk vapor phase molecules interact with bulk liquid/solid phase molecules, exchanging momentum and energy and becoming bulk liquid phase molecules. Condensation from the vapor phase directly to the solid phase (the opposite of sublimation) is possible but not presently of importance in heat-transfer process equipment and will not be discussed further.

Inhomogeneous condensation in heat-transfer process equipment occurs by filmwise or dropwise condensation. Figure 2 shows these two processes schematically. Filmwise condensation occurs when a vapor at saturation temperature comes into contact with a liquid or solid that is below the working fluid saturation temperature. Energy is transferred due to this temperature difference, resulting in vapor condensing into liquid at the interface. Momentum transfer occurs due to two effects, the shear forces produced by the velocity difference between the vapor and the liquid, and convective momentum transport due to momentum formerly carried by the condensed vapor but now residing in the bulk liquid. Energy and momentum are transported from the bulk liquid to the solid due to temperature and velocity differences, respectively.

Superheated vapor must be cooled to saturation conditions in order to condense. This can occur by convective cooling of the bulk vapor without

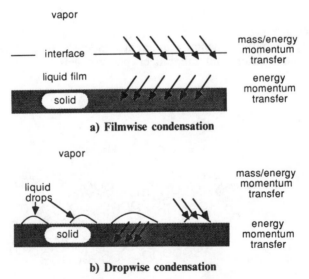

a) **Filmwise condensation**

b) **Dropwise condensation**

Fig. 2   Condensation process schematics.

condensation, wherein the mixed mean temperature of the vapor is brought to saturation conditions, or by cooling of a vapor layer near the liquid to saturation conditions while the bulk vapor core remains superheated.

For both filmwise and dropwise condensation the local temperature difference between the liquid and vapor at the interface is normally small and usually neglected for heat-transfer calculations, although this may not be true for low-Knudsen-number situations when boundary-layer models no longer apply (the Knudsen number is the ratio of molecular mean free path to some characteristic system dimension). Energy transfer away from the condensation interface is controlled by conduction/convection through the associated liquid phase to the cooled solid surface. The condensation heat-transfer coefficient therefore depends on the thickness of the liquid layer. Modeling efforts described later deal basically with calculations of the film thickness.

Ground-based condensing equipment relies principally on gravity to remove the liquid film, e.g., to drain liquids from the outside of condenser tubes, although direct contact spray condensers are also available. Note that both of these rely on gravity to settle the liquid droplets out of the bulk vapor.

Filmwise condensation occurs when the liquid phase wets the solid surface, and dropwise condensation occurs when the liquid phase does not wet the solid surface. Filmwise condensation occurs with most metal surfaces and working fluids. Dropwise condensation usually occurs only for specially treated surfaces, although any nonwetting surface/fluid combination will produce this phenomena. Glass/mercury is such a nonwetting pair discussed in the section Mercury Condensation Experiments.

Condensation of vapor flowing within a constant-diameter tube results in a large decrease in vapor velocity in the direction of flow (as the vapor changes to liquid) due to the density difference between the vapor and liquid phases. This decrease in vapor velocity results in a component of the total pressure gradient that is positive, not negative, in the direction of flow. This is due to the conversion of vapor kinetic energy into vapor static pressure head. This is called pressure recovery. As a result of pressure recovery, the overall pressure gradient may be either positive, negative, or zero. A model for this phenomenon is presented in the section on Pressure Recovery.

The preceding paragraphs have presented a discussion of the basic phenomenology of the condensation processes applicable to thermal energy management and power conversion for space power systems. The following sections describe quantitative models for condensation heat-transfer calculations, beginning with filmwise condensation.

Filmwise Condensation Models

Filmwise condensation within tubes or channels is an important heat-transfer mechanism for many two-phase systems. The following analyses assume that the condensate wets the solid surface (for nonwetting, see the section on Dropwise Condensation). As discussed in the section on

Phenomenology, energy transfer by filmwise condensation is controlled principally by the thickness of the liquid region separating the vapor from the cooled solid surface.

Nusselt-Type Development

Nusselt[4] developed the first filmwise condensation model applicable to a vertical plate. This model has been extended to many geometries and conditions. Figure 3 shows a flat plate at an angle θ to the local acceleration field. The plate is being cooled and the vapor is stagnant. Roshenow and Choi[5] present an analysis for this case neglecting momentum changes in the fluid, interfacial condensation resistance, and interfacial shear. The results for film thickness and average heat-transfer coefficient are

$$\delta(z) = \sqrt[4]{\frac{4k\mu z(T_s - T_w)}{g\,\cos\theta\rho_l(\rho_l - \rho_v)h'_{fg}}} \tag{1}$$

$$h = 0.943 \sqrt[4]{\frac{g\,\cos\theta\rho_l(\rho_l - \rho_v)k^3 h'_{fg}}{L\mu(T_s - T_w)}} \tag{2}$$

where

$$h'_{fg} = h_{fg} + 0.68c_p(T_s - T_w) \tag{3}$$

The h'_{fg} corrects for liquid film subcooling, and L is the plate length.

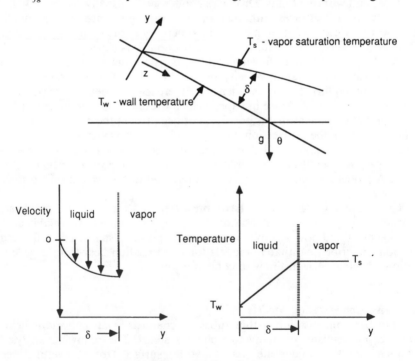

Fig. 3 Flat-plate film condensation.

The film thickness at any position, $\delta(z)$, varies with $g^{-\frac{1}{4}}$, and the overall heat-transfer coefficient varies with $g^{\frac{1}{4}}$. Thus, the heat-transfer coefficient is a slowly varying function of the local acceleration, decreasing with decreasing *g*. In the limit, as *g* goes to 0, the condensate does not drain at all and the stagnant film thickness and heat-transfer coefficient are given by Roshenow[6] as

$$\delta = \sqrt{\frac{2k(T_s - T_w)t}{\rho h_{fg}}} \tag{4}$$

$$h = 1 \bigg/ \sqrt{\frac{2(T_s - T_w)t}{\rho h_{fg} k}} \tag{5}$$

Equations (1) and (2) were derived assuming that the liquid layer is laminar. Alternative derivations would be required if the liquid film were turbulent. However, saturated steam at 1 atm condensing on a vertical surface in a 1-*g* environment would produce a laminar film for a distance of several feet. This distance would be extended for reduced-gravity conditions. Gröber et al.[7] discuss the laminar-to-turbulent transition and the heat-transfer coefficient under turbulent film conditions.

Figure 4 shows a tube cooled internally with condensation on its external surface. The local acceleration is 0, the vapor is stagnant, and film instabilities are neglected. Roshenow[6] gives the film thickness and heat-transfer coefficient as

$$\left[\left(1 + \frac{\delta}{R}\right)^2 + \frac{1}{2}\right]\left[\ln\left(1 + \frac{\delta}{R}\right) - \frac{1}{2}\right] = \frac{2k(T_s - T_w)t}{\rho h_{fg} R^2} \tag{6}$$

$$h = \frac{k}{R \ln\left(1 + \dfrac{\delta}{R}\right)} \tag{7}$$

The film thickness is found using Eq. (6) and then substituted into Eq. (7) to find the heat-transfer coefficient.

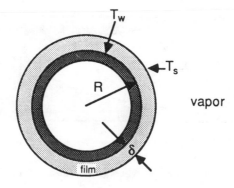

Fig. 4 External film condensation on a cooled tube in 0-*g* stagnant vapor.

Equation pairs (4) and (5) and (6) and (7) could be used to calculate the condensation heat transfer for stagnant liquid films mechanically wiped (removed) from a surface. The wiping cycle time would determine the time over which the equations would be averaged in order to calculate the transient energy removal rate. Alternatively, shear at the vapor-liquid interface can be used to move liquid along the condensing surface as described later.

Shear-Driven Condensation

Figure 5 shows a typical case of shear-driven film condensation within an externally cooled constant-diameter tube under microgravity conditions. Vapor enters the tube at saturation conditions, condenses on the cooled wall, wets the surface, and forms an annular film. The local acceleration field is assumed to be too small to affect either the axial film flow or transverse film flow (pooling). Therefore, the film is azimuthally symmetric and moves in the vapor flow direction. Because of the vapor-liquid density difference, vapor velocity decreases in the flow direction as vapor is condensed to liquid. Momentum is transferred to the film by interfacial shear and condensation. Entrainment of liquid from the film into the vapor core may occur. The flow regime will change from annular to slug, to bubbly, to single-phase liquid as condensation is completed.

Modeling of microgravity condensation is not nearly as advanced as modeling of 1-g condensation, although microgravity condensation modeling tends to be a simplification of 1-g work. Keshock and Sadeghipour[8] compared the 1-g model of Rufer and Kezios[9] and the in-tube condensation models of Bae et al.[10] adapted to a 0-g condition. The model of Bae et al. is based on annular film condensation using the von Kármán velocity profile in the liquid film and an allowance for liquid-vapor slip. Keshock and Sadeghipour assumed that the no-slip condition ($S = 1$) best approximated the 0-g environment and evaluated the model of Bae and co-workers accordingly. Figures 6 and 7 show the results of these comparisons.

Figure 6 shows the calculated quality, x vs tube length for the 1- and 0-g models. Figure 7 shows the heat-transfer coefficients for the same conditions. These figures show that 0-g convective condensation is much reduced

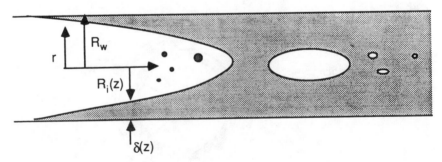

Fig. 5   Microgravity shear-driven film condensation.

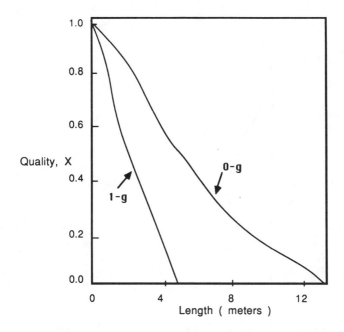

Fig. 6   Quality vs length for 1- and 0-*g* shear-driven condensation.

compared to the 1-*g* model, with 13 m required for complete condensation in 0 *g* vs 5 m for 1 *g*.

The 0- and 1-*g* curves show very different results because of the redistribution of the liquid in the 0-*g* case. Under 1 *g* and for the test conditions shown, the vapor entrance velocity is 1 mm/s and the liquid drains to the bottom of the tube. Although this liquid layer effectively shuts off further condensation in the tube bottom, the drained upper portion of the tube continues to allow condensation. Under 0 *g* and low interfacial shear, the liquid is assumed to be evenly distributed in a thick annular film effectively degrading condensation everywhere. The work of Keshock and Sadeghipour[8] transforms ground-based, gravity-drained condensation to 0-*g* conditions with extremely low interfacial shear. Their results are not characteristic of shear-driven condensation conditions.

Chow and Parish[11] analyzed shear-driven in-tube condensation. They assumed that the liquid film was thin compared to tube diameter, that the film was laminar, that the vapor core velocity profile was flat, and that the system experienced absolute 0-*g* conditions. Under laminar vapor flow conditions, interfacial shear was assumed to be given by the standard laminar friction factor. Under turbulent vapor flow conditions, interfacial shear was assumed to be given by the formulation of Henstock and Hanratty,[12] which gives interfacial shear as a function of the relative properties of the liquid and vapor phases. Momentum changes in the liquid were neglected, as was pressure recovery in the vapor core. Vapor and

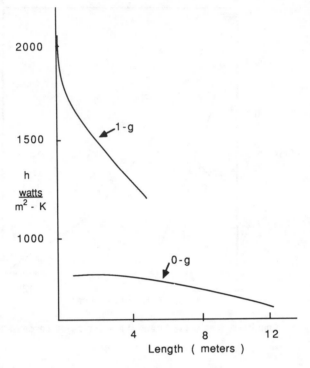

Fig. 7   Heat-transfer coefficient vs length for 1- and 0-g condensation models.

liquid properties were assumed constant at saturation conditions through-out, and entrainment was not considered. The controlling feature of the model, the liquid film thickness, was calculated by carrying out a momentum balance on the liquid layer incorporating momentum transfer to the liquid due to friction and condensation. Energy transfer through the film was assumed to be by planar conduction and similarly for momentum transfer. The velocity of the liquid at the interface, $u_{l\delta}$, is given by

$$u_{l\delta} = \frac{\tau_\delta \delta}{\mu_l} \qquad (8)$$

where δ is the layer thickness, τ_δ the momentum transfer at the interface, and μ_l the liquid viscosity. The film Reynolds number is based on the average liquid velocity and is given by

$$Re_l = \frac{\rho_l \tau_\delta \delta^2}{2\mu_l^2} \qquad (9)$$

Mass conservation dictates that the average vapor velocity at any axial position, u_v, be related to the inlet vapor velocity and Reynolds number by

$$u_v = u_{v0} \frac{Re_v}{Re_{v0}} \qquad (10)$$

The liquid and vapor Reynolds numbers at any axial position are related to the inlet vapor Reynolds number by their respective viscosities:

$$Re_v = Re_{v0}\left(1 - \frac{4Re_l\mu_l}{Re_{v0}\mu_v}\right) \qquad (11)$$

The change in the liquid Reynolds number with axial position is due to condensation which is controlled by conduction through the liquid layer. The increase in the liquid Reynolds number is given by

$$\frac{\mathrm{d}Re_l}{\mathrm{d}z} = \frac{k_l(T_s - T_w)}{\mu_l h_{fg}\delta} \qquad (12)$$

Momentum transfer to the liquid film occurs due to friction and vapor condensation. The net momentum transfer to the film is given by

$$\tau_\delta = \left(\frac{C_f}{2}\right)\rho_v(u_v - u_{l\delta})^2 + \left[\frac{k_l(T_s - T_w)}{h_{fg}\delta}\right](u_v - u_{l\delta}) \qquad (13)$$

where the friction terms are as given by Eq. (14) for laminar vapor and Eqs. (15–17) for turbulent vapor:

$$\frac{C_f}{2} = \frac{8}{Re_v} \quad \text{for} \quad Re_v < 2300 \qquad (14)$$

$$\frac{C_f}{2} = 0.023 Re_v^{-0.2}(1 + 850F) \quad \text{for} \quad Re_v > 2300 \qquad (15)$$

$$F = \frac{\gamma(\mu_l/\mu_v)}{(\rho_l/\rho_v)^{\frac{1}{2}} Re_v^{0.9}} \qquad (16)$$

$$\gamma = ([0.707(4Re_l)^{0.5}]^{2.5} + [0.0379(4Re_l)^{0.9}]^{2.5})^{0.4} \qquad (17)$$

where all variables are as previously defined, $u_{l\delta}$ is the liquid film velocity at the interface, τ_δ the interfacial momentum transport, and u_{v0} and Re_{v0} are, respectively, the vapor velocity and Reynolds number at the tube entrance. The F and γ are formulations given by Henstock and Hanratty.[12]

Chow and Parish[11] solved these equations numerically for steam at 1 atm. The liquid film properties were evaluated at the average film temperature. Table 1 lists the parameters chosen for the evaluation, and Figs. 8 and 9 plot the resulting dimensionless film thickness and vapor Reynolds number as functions of tube length for two vapor inlet Reynolds

Table 1   Convective condensation parameters

	$\Delta T = 15°C$	$\Delta T = 30°C$
T_{sat}	100°C	100°C
μ_l/μ_v	24.92	25.98
ρ_l/ρ_v	1614	1620
$c_p(T_s - T_w)/h_{fg}$	0.02797	0.0559
$Pr_l = c_p\mu_l/k_l$	1.91	1.99

numbers and saturation minus wall temperature differences. Chow and Parish made no comparisons with data.

Chow and Parish point out that film thickness is found to be a small fraction of tube diameter. Therefore, their planar geometry assumption is valid. The figures show that, as expected, increasing the saturation-to-wall temperature difference $T_s - T_w$ from 15°C to 30°C causes the condensation rate to increase, as evidenced by a more rapidly thickening film and decreasing vapor Reynolds number.

The vapor Reynolds number gives an indication of the amount of vapor that remains at a given axial position. Thus, for Re_{v0} of 5000, about 50% of the vapor has been condensed at 8 diam into the tube for a 15°C ΔT. Results shift accordingly for the Re_{v0} of 50, 000 shown in Fig. 9. The heat-transfer coefficient at a given axial position is equal to the liquid conductivity divided by the product of the dimensionless film thickness and the inner diameter. Figures 8 and 9 allow one to evaluate the condensation heat-transfer coefficient, showing that it decreases by orders of magnitude within 10 diam. of the inlet.

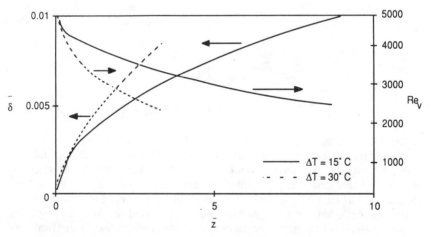

Fig. 8   Dimensionless film thickness and Reynolds number vs length (Chow and Parish[11]).

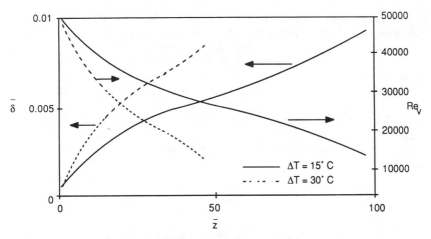

Fig. 9 Dimensionless film thickness and Reynolds number vs length (Chow and Parish[11]).

Although Chow and Parish's model properly accounts for the condensation process thermal energy balance, it does not account for either vapor momentum pressure recovery effects or flow regime transitions which fundamentally alter the postulated phenomena. The rapid vapor deceleration shown by the decrease in the vapor Reynolds number in Figs. 8 and 9 would be accompanied by a rise in the vapor static pressure. This pressure rise could be greater than the frictional pressure drop, thereby resulting in flow reversal/oscillations that would invalidate the proposed condensation model. Even if pressure recovery does not cause flow reversal, the annular flow regime will change to a slug flow condition as condensation proceeds, again changing the condensation geometry and negating the proposed condensation model. Note that, in Eq. (13), momentum transfer to the film due to condensation is assumed to transfer the momentum of vapor moving at the bulk velocity rather than some velocity characteristic of the vapor velocity near the film. Finally, the model is based on true 0 *g* and does not consider body forces of any kind. The effects of spacecraft translational or rotational accelerations on condensation are not addressed. The work of Chow and Parish models the thermal aspects of shear-driven film condensation processes; however, auxiliary analyses must be found to set limits on the range of applicability of the basic model.

Pohner[13] modeled condensing and evaporating annular flows in circular tubes by performing a momentum balance on the liquid film. His analysis is carried out in circular geometry, allowing film thicknesses to grow radially. He assumed that the liquid film remained laminar but that the vapor core could be either laminar or turbulent. Entrainment was not considered, but an explicit expression for the local acceleration was included; therefore, his results may be applied to conditions of reduced gravity (e.g. 1/3, 1/6, 1/100 *g*) both opposed to and in the flow direction. Pressure recovery in the vapor was not considered, nor was vapor superheat.

Interfacial shear was that given by smooth tube relationships. The vapor-liquid interface was assumed to be at saturation conditions. Applying conservation of mass, momentum, and energy, Pohner arrived at the following calculational sequence (all dimensionless variables and relationships are listed in Table 2):

(1) Inlet temperature, pressure, vapor, and liquid mass flow (m_i^*) rates are known (some minimal liquid flow is necessary).

(2) The entrance interfacial shear $[T_i^*(z = 0) = T_{i0}^*]$ and film thickness (actually vapor core radius $[R_i^*(z = 0) = R_{i0}^* = R_{i0}/R_W]$) are found by the simultaneous solution of the following two dimensionless equations for τ_i^* and R_i^*:

$$m_i^* = \frac{\pi \tau_i^*}{4R_i^*}[1 - (R_i^*)^2]^2 + \frac{\pi g_z^*}{8}[1 - (R_i^*)^2]^2$$

$$- \frac{\pi g_z^*}{4}(R_i^*)^2[1 - (R_i^*)^2] - \frac{\pi g_z^*}{2}(R_i^*)^4 \ell n R_i^* \qquad (18)$$

$$\tau_i^* = \frac{c(\mu_v^*)^n(\rho_v^*)^{1-n}f_i^{2-n}}{2^{3+n}(R_i^*)^n} \qquad (19)$$

3) Having fully established the entrance conditions, the following equation for the gradient of the dimensionless vapor core radius may be evaluated:

$$\frac{dR_i^*}{dz^*} = \frac{-(\Delta T)^* f_5}{(\ell n R_i^*)(f_2 f_5 + f_3 f_4)} \qquad (20)$$

4) The value of R_i^* at the next step is found by numerically integrating Eq. (20) (Pohner used Euler's method with a step size equal to half the local film thickness).

5) The dimensionless heat-transfer coefficient is found from

$$h^* = -\frac{1}{\ell n R_i^*} \qquad (21)$$

6) Equation (20) is reevaluated at each spatial step and integrated until the desired condenser exit conditions are achieved.

Figure 10 presents the results of Pohner's calculations for the axial distribution of liquid film thickness, heat-transfer coefficient, average liquid velocity, interfacial velocity, average vapor velocity, interfacial shear stress, and wall shear stress for a cocurrent downward condensing flow of steam at 100 kPa. The quantities have been normalized to their inlet values.

The figure shows the qualitatively expected variation of these quantities. Film thickness grows and the heat-transfer coefficient decreases down the channel. Vapor velocity and interfacial shear decrease down the channel. Liquid film average and interfacial velocities increase down the channel. These quantities change most quickly in the entrance region because the

Table 2 Dimensionless variables and expressions for Pohner model[13]

$$R_i^* = \frac{R_i}{R_W} \qquad \tau_i^* = \frac{\tau_i \rho_l R_W^2}{\mu_l^2} \qquad \dot{m}_l^* = \frac{\dot{m}_l}{\mu_l R_W}$$

$$m_T^* = \frac{m_{\text{Total}}}{\mu_l R_W} \qquad \rho_v^* = \frac{\rho_v}{\rho_l} \qquad \mu_v^* = \frac{\mu_v}{\mu_l}$$

$$\Delta T^* = \frac{2\pi k_l(T_s - T_w)}{\mu_l h_{fg}} \qquad g_z^* = \frac{g_z \rho_l(\rho_l - \rho_v)R_W^3}{\mu_l^2}$$

$$f_1 = \{8m_T^*R_i^* + 2\pi\tau_i^*[(2\rho_v^* - 1)R_i^{*4} + 2(1 - \rho_v^*)R_i^{*2} - 1]$$
$$+ \pi g_z^* R_i^*[(2\rho_v^* - 3)R_i^{*4} + 4(1 - \rho_v^*)R_i^{*2}\ell n R_i^* + 2(\rho_v^* - 2)R_i^{*2} - 1]\}/8\pi\rho_v^* R_i^{*3}$$

$$f_2 = \frac{-\pi T_l^*}{4R_i^{*2}}(1 - R_i^{*2})^2 - \pi T_l^*(1 - R_i^{*2}) - \pi g_z^* R_i^*(1 - R_i^{*2}) - 2\pi g_z^* R_i^{*3}\ell n R_i^*$$

$$f_3 = \frac{\pi(1 - R_i^{*2})^2}{4R_i^*}$$

$$f_4(R_i^*,\tau_i^*) = \pm \frac{(2-n)C\mu_2^{*n}}{\pi 2^{5+n}\rho_2^{*n}R_i^{*4+n}}[\pm f_1(R_i^*,\tau_i^*)]^{1-n}$$

$$\times \frac{\{-8\dot{m}_T^*R_i^* + \pi\tau_i^*[(-1 + 2\rho_2^*)R_i^{*4} - 2(+1 - \rho_2^*)R_i^{*2} + 3]}{+ \pi g_z^* R_i^*[4(1 - \rho_2^*)R_i^{*4}(\ell n R_i^*) - R_i^{*4} + 1]\}}$$

$$\pm \frac{nC\mu_2^{*n}\rho_2^{*1-n}}{2^{3+n}R_i^{*n+1}}[\pm f_1(R_i^*,\tau_i^*)]^{2-n}$$

$$f_5(R_i^*,\tau_i^*) = 1 \pm \frac{(2-n)C\mu_2^{*n}}{2^{5+n}\rho_2^{*n}R_i^{*3+n}}[\pm f_1(R_i^*,\tau_i^*)]^{1-n}$$
$$\times [(-1 + 2\rho_2^*)R_i^{*4} + 2(1 - \rho_2^*)R_i^{*2} - 1]$$

film thickness is thinnest there and the heat-transfer coefficient is greatest, and thus more condensation occurs.

Figure 11 shows the axial variation of the heat-transfer coefficient for cocurrent downward flow of steam in various gravity fields. At 1 m into the tube the heat-transfer coefficient for 0 *g* is one-half the 1-*g* value. This is due to the thicker liquid film that results when the 1-*g* body force pulling liquid down the wall is removed. It is interesting to note that the gravity effect is this large even though the inlet vapor velocity is 3.5 m/s. Recall that from Fig. 7 Keshock and Sadeghipour[8] found that the 0-*g* heat-transfer coefficient was less than one-third the 1-*g* heat-transfer coefficient, even though their calculations were for very-low-vapor Reynolds number.

Figure 12 shows the computed local heat-transfer coefficient plotted vs the position with experimentally measured heat-transfer coefficient data. The condensing experiment is for downward cocurrent condensation of

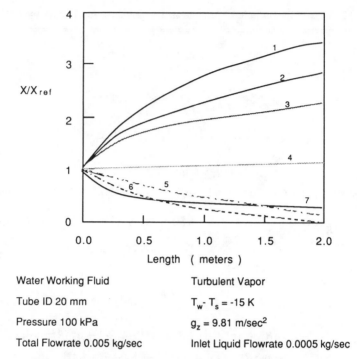

Water Working Fluid   Turbulent Vapor

Tube ID 20 mm $T_w - T_s = -15$ K

Pressure 100 kPa $g_z = 9.81$ m/sec^2

Total Flowrate 0.005 kg/sec   Inlet Liquid Flowrate 0.0005 kg/sec

Fig. 10   Axial variation of normalized film thickness (1), average liquid velocity (2), interfacial velocity (3), wall shear stress (4), average vapor velocity (5), interfacial shear stress (6), and heat-transfer coefficient (7) for cocurrent, downward condensing stream flow.

steam and was carried out by Goodykoontz and Dorsch.[14] Pohner's model predicts the heat-transfer coefficient very well until about 1.5 m into the tube where the model and data diverge. The change in slope of the data at this point may indicate a change in the flow regime occurring in the pipe, invalidating Pohner's model for this region.

Figure 13 compares Goodykoontz and Dorsch's data with Pohner's calculations for mean heat-transfer coefficients in different ranges of vapor inlet Reynolds number. The data and predictions fall within 30% of each other over most of the range, indicating the relative accuracy of the model.

Superheated Vapor Effects

The shear-driven condensation models presented in the preceding section assume that the vapor is at saturation conditions prior to condensation and that the energy transfered during condensation is that of the latent heat of evaporation. Nevertheless, the vapor may have been superheated, in which case a sensible temperature decrease must occur prior to condensation. Eastman et al.[16] present an enthalpy balance analysis to determine the location in a channel at which a superheated vapor would be cooled to

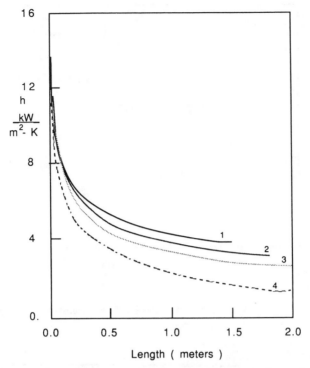

Fig. 11   Heat-transfer coefficient as a function of position for various *g* fields: 1 *g* (1), 0.61 *g* (2), 0.31 *g* (3), 0 *g* (4).

saturation conditions. However, condensation experiments (see the section on Reduced-Gravity Condensation Experiments) have shown that condensation may occur on a cooled pipe wall even though large superheats exist in the flowing vapor core.

One may choose to entirely neglect the superheat energy in comparison with the latent heat of vaporization. Figure 14 is a plot of the ratio of the superheat energy divided by the latent heat of vaporization as a function of superheat temperature for several working fluids at atmospheric pressure. The curves show that the superheat energy may be neglected for less than 50 K superheat in comparison with condensation modeling accuracy. This might not be a reasonable approximation for working fluids with very low heats of vaporization or close to critical pressure conditions.

Dropwise Condensation

Dropwise condensation occurs when the condensing liquid phase does not wet the cooled surface. Heat-transfer coefficients 10 or more times those of filmwise condensation are common. The major heat-transfer effect is due to small drops (100 μ or less for water at 373 K), whereas the large (visible)

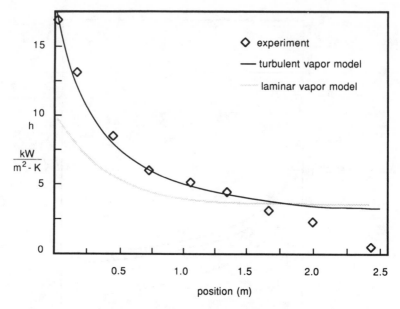

Fig. 12   Computed and experimental local heat-transfer coefficients.[13]

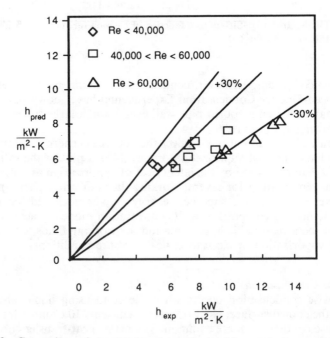

Fig. 13   Comparison of computed and experimental heat-transfer coefficients.[14]

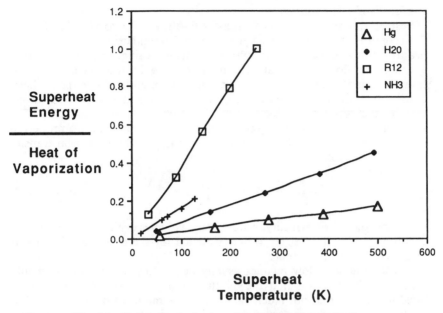

Fig. 14   Ratio of superheat energy to heat of vaporization.

drops are relatively inert. The active, small condensation drops nucleate in cracks and pits on the surface where liquid remains from previous drops (analogous to bubble nuclei for boiling). Small drops grow until they touch other drops. These drops coalesce, continuing to grow by merging with smaller drops. Under microgravity conditions, the large droplet population is depleted by entrainment into the vapor or removed by mechanical means.

Dropwise condensation depends on the large droplet coalescense and removal rate, the effective constriction of the wall conduction heat-transfer path due to highly localized condensation nuclei, the resistance of a drop promoting coating (if any), and noncondensable gases. No generally applicable quantitative heat-transfer models have been developed, although empirical relationships do exist.[19] Dropwise condensation microgravity experiments are discussed in the section on Mercury Condensation Experiments.

Bulk Condensation

Microgravity film or dropwise condensation occurring in a channel produces an increasing liquid fraction until the flow regime changes to slug or bubbly flow. This flow regime transition alters the heat-transfer mechanism. No models for slug flow condensation in a microgravity environment have been developed. No data have been explicitly identified as slug flow condensation, although these conditions must have existed in many experiments.

Bubbly Flow Condensation

Best et al.[20] proposed a heat-transfer coefficient for low-quality bubbly flow conditions. They assumed that the wall boundary-layer resistance to heat transfer was the same for bubbly flow as for a single-phase flow having a volumetric flow rate equal to that of the bubbly flow. Therefore, heat transfer is limited by energy transport through an all-liquid boundary layer at the tube wall, and the velocity profile in this boundary layer is characterized by the superficial velocity of the two-phase vapor-liquid core. The vapor and liquid velocities were assumed to be equal, and all properties were evaluated at bulk fluid saturated conditions. The local heat-transfer coefficient is given by

$$\frac{hd}{k} = 0.023 Re^{0.8} Pr^{0.3} \left(1 + x \frac{v_{fg}}{v_f} \right)^{0.8} \quad (22)$$

where x is the flowing quality, v are specific volumes evaluated at saturation conditions, and all other terms are evaluated for the total mass flowing as saturated liquid in the tube.

Bubbly and slug flow regimes usually occur only for low-vapor-quality conditions. This means that the mixture enthalpy is almost that of saturated liquid and that not much vapor energy remains to be removed. Thus bubbly and slug flow regimes only exist in short sections of condenser tubing, and film condensation models are frequently projected to zero-quality ($x = 0$) conditions to approximate this region. No data have been explicitly identified as bubbly flow condensation, although these conditions must have existed in many experiments.

Pressure Recovery

The toal pressure gradient in a steady-state condensing mixture is composed of a frictional pressure gradient and a momentum or spatial acceleration component. The velocity of a two-phase mixture decreases in the flow direction due to the density change of the condensation process. The kinetic head of the mixture is converted to static pressure head. This kinetic conversion is a pressure gain as opposed to the frictional pressure loss. The frictional gradient and the spatial acceleration gradient are of opposite sign, and the net effect of the two can produce either a static pressure gain or loss in a given system. The spatial acceleration term can be written as a general function of mass flux, quality, and void fraction, as shown by

$$\left(\frac{dp}{dz} \right)_a = -G^2 \frac{d}{dz} \left[\frac{x^2 v_g}{\alpha} + \frac{(1-x)^2 v_f}{(1-\alpha)} \right] \quad (23)$$

Calculation of the frictional component depends on the model for a particular flow regime and must be combined with the spatial acceleration to compute the net pressure gradient. Very high condensation rates produce large pressure recoveries that can make the overall pressure gradient sufficiently positive to cause flow instability.

Noncondensable Gas Effects

Figure 15 shows the geometry of in-tube condensation of a working fluid vapor mixed with a noncondensable gas phase. Condensation removes working fluid vapor from the mixture, but not the noncondensable gas. This results in noncondensable gas buildup in both the radial and axial directions.

Working fluid vapor must diffuse and convect radially through an increasing noncondensable gas concentration in order to reach the gas-liquid interface and condense. This results in an increased resistance to condensation which can dominate the overall heat-transfer coefficient for very small amounts of noncondensable gas in the inlet stream. Collier[17] has analyzed this case for simple systems.

The noncondensable gas layer is also moved axially in the direction of flow under the influence of shear force from the mixture's core region. The noncondensables concentrate at the axial position of the gas mixture-liquid interface, where entrainment of the noncondensable gas into the liquid takes place.

The type of entrainment mechanism depends on whether filmwise or dropwise condensation is occurring. For filmwise condensation, the non-condensable gas is thought to be entrained at the transition from annular to slug flow. At this point the liquid layer forms an azimuthal wave that bridges the channel and entraps a volume of gas. This gas is then carried away by the liquid phase.

For dropwise condensation, Lancet et al.[18] observed the buildup of a noncondensable gas plug upstream of the liquid interface. Entrained liquid droplets visibly slowed as they entered the gas plug region. The droplets collected there until they formed a liquid slug that filled the tube and carried the noncondensable gas away.

Buoyancy can affect microgravity mixed-component condensation if the velocity is sufficiently low and the acceleration nonzero. However, for most shear-driven condensation this is not the case and 1-*g* models should be acceptable (although this remains to be proven).

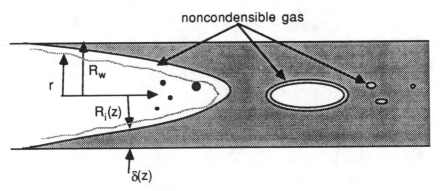

Fig. 15 Noncondensable gas buildup in shear-driven condensation.

Reduced-Gravity Condensation Experiments

Most reduced-gravity two-phase flow experiments have dealt with boiling or with adiabatic flow regime and pressure drop measurements. Of the condensation experiments that have been conducted, the major emphasis has been on demonstrating component feasibility or operational stability. The following paragraphs describe microgravity condensation experiments conducted using mercury, Freon, and water as the working fluids.

Mercury Condensation Experiments

Interest in microgravity liquid metal two-phase flow was stimulated in the late 1950's and early 1960's by plans to use liquid metal Rankine cycle power systems for spacecraft. Since so little reduced-gravity two-phase flow information was available, early tests tended to be concept demonstrations and scoping experiments rather than tests designed to quantitatively study the mercury microgravity condensation process.

Reitz[21] studied the stability of microgravity condensing mercury within glass tubes using the "loop" shown in Fig. 16. A pool-type boiler was used to produce saturated mercury vapor at 533 K. A 100-mesh/in. screen hindered liquid escape from the boiler during 0-g periods. The mercury vapor was used at saturation conditions or superheated to 644 K. Condensation was observed in a glass tube with a 5-mm i.d. and a 0.305-m length. Cooling was provided by dry-ice-cooled air blown over the outside of the glass condenser tube. The valve shown in the figure was shut during the 0-g parabola, and condensate was allowed to build up at the condenser end away from the boiler. Inlet vapor velocity was on the order of 30 m/s. A typical parabola resulted in approximately 13 mm of Hg collecting in the end of the condenser. The return valve was opened at normal gravity, and the condensate was allowed to flow back to the boiler by gravity. Reduced gravity was produced by flying parabolic trajectories in an Air Force C-131B aircraft. The reduced-gravity periods lasted for 11–15 s.

Retiz[21] observed dropwise condensation and a stable vapor-liquid interface that developed in the tube for both ground-based and 0-g conditions.

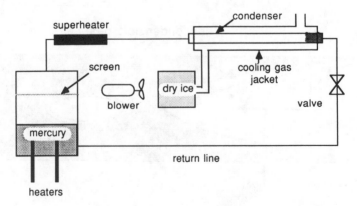

Fig. 16   Mercury condensing test apparatus of Reitz.[21]

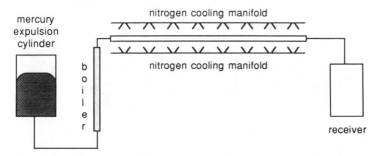

Under 0 *g*, condensate droplets were uniformly distributed around the tube surface but tended to collect in the tube bottom under normal-*g* conditions. It was observed that droplet size was inversely related to both the condensing rate and mercury gas velocity. No quantitative measurements were made that would allow the local heat-transfer coefficient to be determined. The vapor-liquid interface was experimentally found to be stable at normal gravity for tubes with inner diameters up to 3.8 mm.

Small but unquantified amounts of noncondensable gases were reported to have significant effects on system operation. Noncondensables collected at the vapor-liquid interface and tended to slow mercury droplets by drag before the droplets could impact on the interface. The droplets tended to collect at the gas-vapor interface and form a liquid slug at that location. The noncondensable gas bubble and associated liquid slug reduced the condenser tube length available for condensation, degrading overall system performance. Summarizing the Reitz experiment, the feasibility of in-tube non-wetting mercury condensation under 0-*g* conditions was experimentally demonstrated, and recommendations were made as to the importance of removing noncondensable gases from such systems.

The NASA Lewis Research Center conducted a series of tests on reduced-gravity mercury condensation in the early 1960's. The results of these tests are documented in reports[22-26] and films.[27, 28] The test objectives were to observe stable 0-*g* mercury condensation, to measure the effect of gravity on the condenser tube frictional pressure drop, to study tapered vs constant-diameter tube performance, and to estimate the condensation heat-transfer coefficient. Figure 17 shows a simplified schematic of the mercury test system.

The test system consisted of a gas pressurized bladdered expulsion cylinder, a perheater-high flux boiler-main boiler, a vapor flow venturi, a condensing tube, and a receiver. Condenser cooling was provided by gaseous nitrogen in crossflow, except for one test series that used a NaK counterflow tube-in-tube arrangement. Thermocouples and inductance-type pressure transducers were used to measure test conditions. The system was used in ground and 0-*g* parabola testing conducted in the bomb bay of an AJ-2 aircraft. Zero-gravity periods lasted about 10–15 s. Table 3 lists the test conditions. Note that Pyrex, steel, copper, and 9M tubes were used.

Table 3   Range of variables for mercury condensing experiments

Item	Crossflow: Nitrogen-cooled					Counterflow: NaK-cooled
	Gravity comparison tests			Wetted comparison tests		
Gravity level, g	1, 0	1, 0	1, 0	1	1	1
Wetted condition	Nonwetted	Nonwetted	Nonwetted	Nonwetted	Wetted	Nonwetted
Tube material	Pyrex	Stainless steel 304	316 Stainless steel	304 Stainless steel	Copper	9M
Tube diameter, mm	Constant 6.9, 10.2, 12.4	Constant 7.9	Tapered 10.2–3.8	Tapered 12.7–5.1	Tapered 12.4–5.1	Tapered 11.4–5.1
Condensing length, m	1.73[a]	1.14–1.83	1.14–1.83	0.46–1.14	0.46–1.09	0.18–1.07
Vapor mass flow rate, g/s	11.4–22.7	11.4–22.7	11.4–22.7	11.4–22.7	11.4–22.7	18.6–20.5
Vapor inlet quality	0.85–1.0	0.85–1.0	0.85–1.0	0.80–1.0	0.80–1.0	1.0
Vapor inlet pressure, kPa	82.7–179	110–152	82.7–152	89.6–124	96.5–131	103–152
Vapor inlet velocity, m/s	36.6–116	5.8–101	36.6–70.1	27.4–52.9	27.4–27.9	36.6–45.7
Vapor inlet Reynolds number $\times 10^{-4}$	2.0–5.5	3.0–5.0	2.0–4.0	1.5–4.0	1.5–4.0	3.0–4.0

[a]Single value.

Constant-diameter and tapered-diameter tubes were tested. Both mercury wetting and nonwetting materials were employed.

The steps in a typical run were as follows: The system was evacuated to a pressure of 60 μ of mercury, and the heaters were brought up to operating temperature. Mercury vapor was allowed to flow through the system for 5 min to remove noncondensable gases. The receiver pressure was adjusted to a constant value between 96.5 and 103 kPa, and the nitrogen gas flow adjusted so that condensation produced a mercury vapor-liquid interface at the desired location. The 1- and 0-*g* tests were accomplished by recording the 1-*g* data points while the aircraft was in level flight and then immediately after, in 0 *g*, recording the 0-*g* data points without changing system operating conditions.

The complete test data are tabulated in the final reports. These tables list *g* level, parabola identification number, the condensing length (distance from tube inlet to vapor-liquid interface), liquid and vapor inlet mass flow rates, inlet quality, the static pressure at six locations along the condenser, and the temperature of the entering mercury mixture. Observations of flow

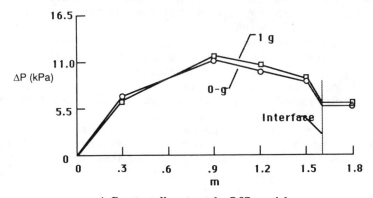

a) **Constant-diameter tube 7.87 mm i.d.**

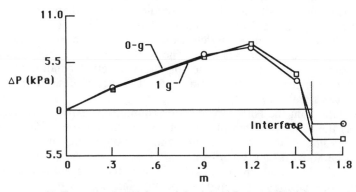

b) **Tapered tube 10.2 mm inlet i.d., 4.7 mm outlet i.d.**

Fig. 18   Mercury condensing flow static pressure drop for 17.3-g/s vapor inlet flow.

regime and droplet characteristics are presented in written form as well as films. A summary of the overall results follows in the following paragraphs.

The tests showed that mercury condensing in tubes can operate stably in 1 and 0 g for the conditions tested. A superheated mercury vapor core was inferred to exist throughout most of the axial vapor length. Droplet condensation began a short distance after the vapor entered the condensing tube, and droplets on the wall were dragged in the flow direction by vapor shear, agglomerating with other droplets as they moved. Droplets were entrained in the vapor stream upon reaching a sufficiently large size. Droplets tended to collect in the bottom of the tube under 1-g conditions. The droplet size distribution on the wall was independent of gravity. A stable vapor-liquid interface formed in both 1- and 0-g tests. The interface inclined under the influence of gravity, but was normal to the tube axis in 0 g. Two flow regimes were reported to exist: droplet (dispersed or fog) flow and a single-phase liquid filling the tube. Flow regime observations were not made for the wetting flow of mercury in copper tubing. Figure 18 shows a plot of the local static pressure drop as function of position for tapered and constant-diameter tubes under 1- and 0-g conditions. Measurement accuracy was not explicitly reported; however, the 1- and 0-g curves are described as indistiguishable.

The gaseous nitrogen crossflow cooling is thought to have produced a nearly linear decrease in quality as a function of position. Therefore, the pressure gradient due to vapor condensation momentum recovery is positive and approximately constant. The frictional pressure gradient varies with the mixture velocity and tube diameter and is largest at the inlet, but decreases with position, becoming smaller in magnitude than the pressure recovery after about one-half and three-fourths the tube length for the constant-diameter tube and tapered tubes, respectively. For the range of flow rates and condensing lengths tested, the overall static pressure difference varied between a pressure rise of 6.2 kPa and pressure drop of 0.7 kPa.

The pressure drop data for the nonwetting tests were compared with the Lockhart-Martinelli and "fog" flow[29] models. The Lockhart-Martinelli correlation agreed with the data for qualities greater than 0.4 but underpredicted the pressure drop by a factor of 10 at the lowest qualities tested. The fog flow model roughly matched the data over the whole quality range, but considerable scatter was seen. The wetted test data showed pressure drops slightly greater than those for the nonwetted condition, but the overall trends remained the same as for the wetted case, and the data scatter was significant.

The data from the NaK-tapered tube counterflow experiment were used to study the condensation heat-transfer coefficient. Because of limitations in the experimental setup, the mercury condensation coefficient could only be said to be above 17.5 kW/(m²-K).

Considered in total, the NASA Lewis mercury tests showed that stable mercury condensing tubes could be operated in a 0-g environment and that their performance was independent of gravity. These conclusions only apply for the flow conditions tested.

For example, Table 3 shows that inlet vapor velocities were always in excess of 30 m/s or more. At these velocities friction and inertial effects dominate the 1-*g* gravity effects; thus, it should be expected that the measured pressure drop does not change in 0 *g*. Furthermore, the Lockhart-Martinelli correlation is best applied under adiabatic separated flow conditions rather than the condensing flow conditions of these tests in which saturated liquid droplets are thought to be entrained in a superheated vapor core. Finally, the frictional pressure gradient was extracted from the measured static pressure gradient data by subtracting a calculated momentum recovery. The momentum recovery calculation was based on assuming that the vapor and liquid moved at the same velocity, when, in fact, there is a large velocity difference between the vapor velocity and the droplets on the wall. The friction pressure gradient presented in the plots depends on the validity of this approximation.

Interest in mercury as a working fluid declined with the reduction in the space nuclear power program in the late 1960's. However, interest in microgravity two-phase flow continued with other working fluids being used in thermal energy transport systems. The following section describes work done in Freon experiments.

Freon Experiments

Williams et al.[30] carried out reduced-gravity condensing experiments using R-12 as the working fluid. The test system shown in Fig. 19 was flown in a C-135 aircraft to produce reduced-gravity conditions. The tests were primarily conducted as flow visualization and stability experiments, and no quantitative heat-transfer or pressure drop data comparing 1- and 0-*g* operation were collected.

As the schematic shows, R-12 was stored in a tank that was heated to produce the desired Freon pressure conditions, typically 1520 kPa. The Freon was completely vaporized in an electrically powered helical heater. It then flowed through a receiver tank that damped out possible pressure fluctuations. Freon vapor left the receiver at 1520 kPa and 366 K. It then flowed into a condenser with three parallel tubes made of fused quartz, 2.44 m long and 3.18 mm i.d. It went overboard from the condenser. The quartz tubes were cooled by water flowing parallel to the R-12 in a channal made of 1.27-cm-thick Plexiglas. Cooling water was discharged overboard. High-speed cameras recorded flow regime conditions. Table 4 lists the test conditions. A nine-tube bank of parallel tubes was also flow in the aircraft but is not discussed here because it was not a heat-transfer experiment.

The high-speed photography (3300 frames/s typical) showed annular to slug to bubble condensation flow regimes. Williams[30] reported a "considerable" amount of wave action on the liquid film from the earliest observable point. The film was observed to develop large waves that ultimately bridged the tube, isolating large vapor bubbles. These bubbles then flowed steadily down the tube, decreasing in size until they disappeared. The flow was steady in appearance with no sharp boundary apparent between the single-phase and two-phase condensing regions. Minor oscillations in the

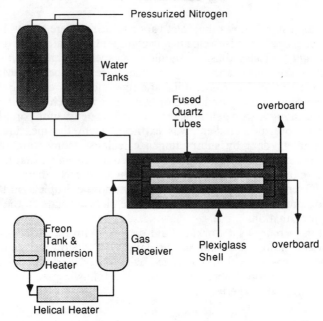

Fig. 19   R-12 condensing apparatus of Williams et al.[30]

liquid leg were seen. The 0-g flows appeared to be more annular in nature than the 1-g flows, with a noticeably smaller liquid film on the tube bottom. Bubbles in the flow tended to be centered in the tube rather than riding along the tube top as they did in 1 g. The 0-g condensation length appeared somewhat shorter than the 1-g condensation length.

This series of experiments may be termed a separate effects test, that is, a high quality source supplied vapor to a condensing section that discharged overboard. There was little opportunity for instabilities to develop based on coupled effects between the boiler and the condenser. Under these conditions, the filmwise condensation process was seen to be hydrodynamically stable in 0 g, an important conclusion.

Table 4   Test conditions for R-12 experiments of Williams[30]

Condensing Freon: tube 3.18 mm i.d., 2.44 m length	
Inlet pressure	1756 kPa
Inlet temperature	377 K
Flow (per tube)	0.63–2.52 g/s
Cooling water	
Pressure	35 kPa
Flow (total)	12.6 g/s
Temperature	not reported but presumed to be ambient

The qualitative observation that the 0-*g* condensation length was shorter than the 1-*g* condensation length is interesting because it is the opposite of the expected result. All of the filmwise condensation models discussed in the section on Filmwise Condensation, predict that 0-*g* condensation should require longer condenser lengths than 1-*g* condensation. However, it is not possible to quantitatively compare the 1- and 0-*g* condensation lengths since the experiment was not instrumented for this purpose.

Before one leaves the Freon experiments section, the Sundstrand work must be mentioned. The Sundstrand Corporation has conducted a set of reduced-gravity flow regime and pressure drop experiments[31,32] using Freon (R-114) in a closed-loop boiling, condensing, and adiabatic test system. The system utilized a swirl flow evaporator and shear-driven condenser. However, the condenser was not instrumented for detailed heat-transfer measurements; and hence, no quantitative condensation modeling information was obtained.

Water Experiments

Feldmanis[33] carried out reduced-gravity two-phase flow experiments using water as the working fluid with the system shown schematically in Fig. 20. The boiler condenser system was free floated within a KC-135 aircraft to produce reduced-gravity periods of 8–19 s. A straight tube electrically heated evaporator 0.91 m long, 9.53 mm o.d., 0.889 wall thickness, and water-cooled tapered condenser were used. Table 5 lists some of the test conditions for this experiment. The test package was not instrumented to measure flows; thus, no quantitative conclusions can be reached regarding the reduced-gravity condensation process. Temperature and pressure data were monitored via a recording oscillograph. Figure 21 is reproduced from Ref. 33 and shows a typical pressure and temperature transient for the condenser inlet. Condenser outlet pressure was noted as differing very little from inlet pressure and was not reported quantitatively.

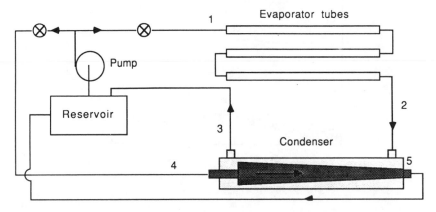

Fig. 20   Water-cooling apparatus of Feldmanis.[33]

Table 5   Test conditions for water experiments of Feldmanis[33]

Condensing water

Inner tapered tube 0.406 m/length; at inlet; 9.53 mm i.d. within a tube having a
  19.1 mm i.d.; at outlet 17.4 mm i.d. within the same 19.1-mm-i.d. tube
Inlet pressure 5.17 kPa (gage); aircraft cabin pressure not reported
Inlet temperature 375 K
Flow 1.01–1.26 g/s (approximate)

Cooling water
Inlet temperature 314 K
Outlet temperature 321 K
Flow not measured

As shown in Fig. 21, Feldmanis[33] reported a system pressure increase
during the reduced-gravity condition. Pressure oscillations in the boiling/
condensing loop were observed before and during the early part of the
parabola. These oscillations damped out 5 s into the reduced-gravity condition, but resumed after the 0-*g* period. Not shown in Fig. 21 but reported
by Feldmanis was a similar pressure variation occurring in the single-phase
cooling loop. The boiler inlet thermocouple was reported to have given
erroneous readings.

Feldmanis attributed the pressure rise in the two-phase portion of the
loop to a change in the boiling heat transfer upon entering 0 *g*, leading to
increased vapor generation and pressure oscillations. The pressure decrease
in the single-phase loop upon ending 0 *g* was attributed to the impact of the
package on the aircraft deck.

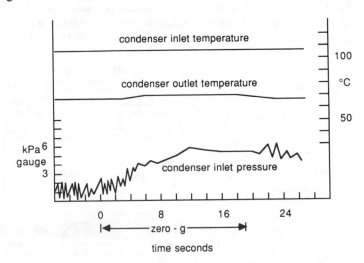

Fig. 21   Condenser inlet and outlet temperatures and inlet pressure from Feldmanis.[33]

Feldmanis did not address the potential effect of the 2-*g* pull-out before and after the 0-*g* period on system performance. Water entered the evaporator at 0.027 m/s. At this low velocity the two-phase mixture could have been in a stratified flow condition during the 2-*g* periods. This would cause steam blanketing of the upper portion of the channel with a resulting wall temperature increase. Upon entering 0 *g*, the liquid would redistribute around the channel and encounter the overheated upper portion of the channel. This could have resulted in the observed pressure fluctuations, which then damped out as the wall was cooled to normal conditions. In addition, components of the 2-*g* acceleration could have contributed to a flow stagnation that was not observable since flow was not being monitored. Alternatively, the 2-*g* pull-out could have caused the pump to lose prime, leading to a pressure drop in both the single-phase and two-phase systems.

As already mentioned, the condensing heat-transfer coefficient was not measured. However, Feldmanis reported that the condensate outlet temperature increased approximately 3 K during the reduced-gravity period. This was attributed to a decrease in the condensation heat transfer since the inlet temperature had not changed. At the low velocity of these experiments, the drainage-driven model of Keshock and Sadeghipour[8] discussed in the section on Shear-Driven Condensation should apply, and Feldmanis' result is consistent with the trend of this model.

Lastly, Feldmanis pointed out, as had others, that when condensation occurs in a tube, the vapor velocity and the interfacial shear stress will decrease. This causes a decrease in condensate film velocity and an increase in film thickness. The increasing film thickness causes a reduced condensation heat-transfer coefficient. Feldmanis recommended a tapered condenser to decrease vapor flow area, thereby maintaining vapor velocity as high as possible and film thickness low. However, as opposed to other researchers, he used an annular condensing space (see Fig. 20) whose inner vapor diameter increased in the flow direction. This had the effect of simultaneously decreasing the vapor flow area and increasing the wetted perimeter, thereby decreasing the film thickness. Unfortunately, no quantitative measurements demonstrate the usefulness of this approach.

Lee[34] and Best and Kachnik[35] conducted flow regime, pressure drop, boiling, and condensation experiments for NASA in the KC-135 aircraft. A schematic diagram of the experimental system is shown in Fig. 22. The basic system consists of two boilers, a water-cooled condenser, a circulating pump, and a water storage tank. The condensing test sections were made of Pyrex, 6 mm i.d., 8 mm o.d., and 0.75 m total length. The primary boiler used in this experiment was supplied by Pacific Northwest Laboratories (PNL). This boiler had transparent boiling sections of 8-mm-i.d., 10-mm-o.d. quartz tubing connected by Teflon T-shaped tube fittings. Flow conditions throughout the boiler and condenser could be visually observed since both were transparent. The other boiler used in the experiment was built by Texas A&M University (TAMU). This boiler consisted of a stainless steel tube, around which was wrapped Fiberglas tape, nichrome wire, and a layer of thermal insulation. The stainless steel tube had a

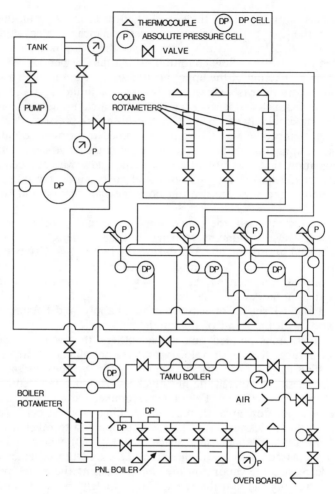

Fig. 22   Lee's water boiling and condensing system.

6.3-mm i.d. and a 9.5-mm o.d. A subatmospheric pressure water and steam mixture was chosen as the working fluid to ensure safe operation of the system in the airplane. The volumetric flow rate was controlled in the range of 2.75–0.005 l/min so that the quality could be varied from 1 to 0. Boiler power was controlled between 0 and 6 kW. Close-up films of the boiling and condensing flows were taken with a high-speed (2700 frames/s) 16-mm motion picture camera. The high-speed camera took films every third parabola. An IBM XT was used to monitor and record temperatures, pressures, and power. A total of 100 parabolas was achieved in four days of flying. Many unexpected two-phase flow phenomena observed were thought to be due to the microgravity simulation method by airplane and not characteristic of reduced gravity per se. Some of these effects are

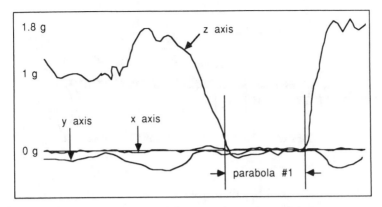

Fig. 23   Typical KC-135 three-axis accelerometer trace.

described below, followed by the condensation heat-transfer results (also
see Chapter 6).

The microgravity conditions were produced by a NASA KC-135 test
aircraft flying from Ellington Field, Johnson Space Center. A typical trace
of the onboard three-axis accelerometer is shown in Fig. 23. Since the
aircraft rotates very slowly about its transverse axis during the parabola,
the forces due to the rotation will be neglected. It was observed that
free-floating objects drift aft perhaps as much as 7 m in 20 s. This says that
the aircraft actually has a mismatch in horizontal velocity between plane
and package of approximately 1.2 km/h. This is an extremely small mismatch, considering that the plane is moving at over 644 km/h. If the
velocity mismatch of 1.2 km/h were due to a constant acceleration, then the
horizontal acceleration would have a value of 0.00155 g. This is the

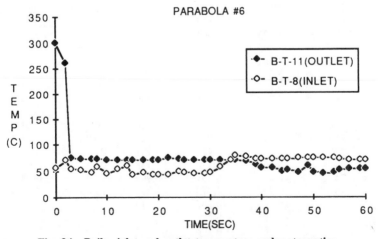

Fig. 24   Boiler inlet and outlet temperature under stagnation.

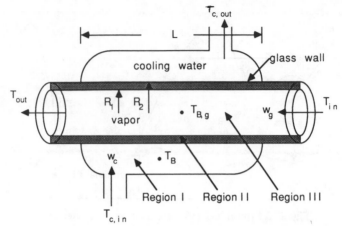

Fig. 25 Condenser section (one of three) schematic.[35]

Table 6 Cooling water average differential temperature, °C

Parabola	CJ1	CJ2	CJ3	Bulk temp., °C	Power, W
1	4.944	4.141	3.705	17.78	1688.2
2	3.045	2.957	2.590	17.94	1054.9
3	2.971	2.971	2.562	18.10	1010.6
4	3.335	3.201	2.761	18.26	1108.1
5	3.516	3.337	2.823	18.42	1035.7
6	4.790	4.336	4.069	18.58	1720.9
7	4.664	4.524	4.069	18.74	1720.8
8	3.677	3.506	3.374	18.90	1571.7
9	1.442	1.688	1.153	19.06	342.8
10	2.354	1.592	0.951	19.22	343.3
11	3.177	2.067	1.506	19.38	186.7
12	4.520	4.527	3.967	19.52	908.3
13	4.773	5.004	4.995	19.68	927.5
14	4.995	5.221	5.333	19.84	1012.3
15	4.161	4.207	3.907	20.00	931.8
16	5.033	5.282	5.403	20.16	1117.7
17	5.406	5.307	5.278	20.32	1083.5
18	4.466	4.516	4.589	20.48	728.7
19	4.115	4.249	3.484	20.64	835.9
20	4.141	4.324	3.640	20.80	885.2
21	3.471	3.099	2.566	20.96	618.9
22	3.893	3.918	3.281	21.12	685.1
23	4.365	4.390	3.738	21.28	666.8
24	6.453	6.751	6.945	21.44	1452.9
25	−0.255	−0.073	−0.293	21.60	775.1
26	−0.364	−0.242	−0.410	21.76	751.2
27	−0.398	−0.280	−0.476	21.92	767.2
28	−0.469	−0.268	−0.536	22.08	1828.6

experimentally observed horizontal acceleration of the KC-135 aircraft, made possible by the pilots flying the aircraft on the basis of the position of the free-floating package and not on an accelerometer indication. In reality, an acceleration of $1.5 \times 10^{-3}\,g$ is below the aircraft accelerometer's detectability.

The plane's interior does seem to have a good vertical 0 g, as seen by the fact that things float up as much as down. Therefore, the plane is flying a ballistic trajectory that compensates very closely for aerodynamic drag.

The 2-*g* pull-out period before and after each 0-*g* period has a profound effect on system operation, such that during many parabolas the system did not reach steady-state conditions. The effect of the 2-*g* pull-out observed during an experiment run is discussed and analyzed in the following paragraph.

Figure 24 is a plot of boiler inlet and outlet temperatures for a 0-*g* period that begins at time 0 and ends at time 25. As shown in the figure, the boiler inlet temperatures were sometimes higher than the feed water temperature

Table 7   Condenser average temperature, °C

Parabola	C-T-8	C-T-9	C-T-10	C-T-11
1	93.08	88.14	81.79	79.22
2	76.22	73.51	71.18	68.23
3	75.09	72.57	70.51	67.79
4	83.95	77.14	72.64	70.12
5	87.12	82.62	79.12	74.86
6	96.41	93.30	89.20	85.70
7	96.43	93.24	89.83	85.93
8	88.16	84.10	81.07	76.42
9	62.66	58.89	54.09	49.74
10	79.42	67.90	58.83	47.86
11	86.50	78.65	63.52	60.32
12	87.05	86.92	84.06	73.58
13	89.69	89.90	88.94	89.23
14	90.43	90.74	90.07	90.06
15	84.73	84.62	82.54	78.65
16	91.78	92.04	91.27	91.04
17	92.41	92.74	91.99	91.96
18	84.55	85.20	84.33	84.44
19	87.65	86.89	83.35	77.40
20	89.65	88.70	85.69	78.46
21	84.04	78.90	71.79	66.89
22	87.06	86.15	82.23	76.03
23	88.39	87.63	84.70	82.82
24	103.59	103.27	102.47	104.08
25	30.80	31.89	29.62	28.83
26	28.78	30.09	27.42	26.90
27	27.66	29.24	25.78	25.54
28	25.91	26.74	26.56	26.19

(25°C) and sometimes even higher than the boiler outlet temperature. This was interpreted as being due to the fact that the boiler tube was positioned in the flight direction during this parabola, and residual acceleration caused hot steam to flow back to the boiler inlet due to buoyancy forces during the 2-g pull-out. Stratification of the two-phase flow during pull-out was observed for low flow test cases. This would lead to local overheating of the boiler wall and high superheating until quenching occurred during 0 g. Despite these artifacts of the test platform, useful data were recorded.

A schematic diagram of a condenser test section is shown in Fig. 25. Temperature and pressure of the two-phase mixture were measured at the inlet and outlet of each condenser section. Single-phase liquid flow at the evaporator inlet was measured with a turbine flowmeter. The absolute and differential temperatures of the single-phase cooling water across each jacket were measured. The cooling jacket flows were set using rotameters during level flight, and a positive displacement pump was used to keep these flows constant during the 0-g period.

Table 8   Volumetric flow rate l/min

Parabola	Cooling jacket 1	Cooling jacket 2	Cooling jacket 3	Feed water
1	0.757	0.757	0.757	0.91
2	0.757	0.757	0.757	0.91
3	0.757	0.757	0.757	0.91
4	0.757	0.757	0.757	0.91
5	0.757	0.757	0.757	0.91
6	0.757	0.757	0.757	0.91
7	0.757	0.757	0.757	0.91
8	0.757	0.757	0.757	0.91
9	0.757	0.757	0.757	0.91
10	0.757	0.757	0.757	0.30
11	0.757	0.757	0.757	0.30
12	0.757	0.757	0.757	0.30
13	0.757	0.757	0.757	0.30
14	0.757	0.757	0.757	0.30
15	0.757	0.757	0.757	0.30
16	0.757	0.757	0.757	0.45
17	0.757	0.757	0.757	0.45
18	0.757	0.757	0.757	0.45
19	0.757	0.757	0.757	0.45
20	0.757	0.757	0.757	0.45
21	0.757	0.757	0.757	0.45
22	0.757	0.757	0.757	0.45
23	0.757	0.757	0.757	0.45
24	0.757	0.757	0.757	0.45
25	0.757	0.757	0.757	0.45
26	0.757	0.757	0.757	0.45
27	0.757	0.757	0.757	0.45
28	0.757	0.757	0.757	1.2

Table 9   Lee and Best's[35] condensation heat-transfer coefficients

Parabola number	Jacket 1, kW/(m²-K)	Jacket 2, kW/(m²-K)
13	2.06	3.03
14	2.67	3.00
15	1.82	1.85

Estimated experimental error ± 15%.

Tables 6–8 list the measurements taken during a day's flying. Table 6 lists the differential temperature rise of the single-phase cooling water averaged over the 0-*g* period, the average temperature of the cooling water, and the electrical power supplied to the boiler. Table 7 lists the two-phase condensing flow temperature at the inlet to the condenser, between each section, and at the outlet. Table 8 lists the flow rates to each jacket and the boiler. This information allows the calculation of condensation heat transfer coefficients in each condenser section.

Examining the data of Tables 6–8 reveals some inconsistencies. For example, the cooling jacket differential temperatures for parabolas 25–28 are negative. There is no known reason for this. However, rather than eliminate data, it is all included so that other researchers may interpret the experiment for themselves. Condensation heat-transfer coefficients were calculated for cooling jackets 1 and 2 for parabolas 13–15 because these parabolas seemed to have understandable thermodynamic conditions. Table 9 lists the measured heat-transfer coefficients. The experimental error associated with these values is estimated to be 15%.

Model Intercomparisons and Validation
The following sections describe some of the validation and comparison available among microgravity condensation models.

Filmwise Condensation
Figure 26 is a plot of quality vs position computed from Pohner's filmwise condensation model[13] and Chow and Parish's[11] filmwise condensation model for the conditions of parabola 15 from Lee and Best's experiments.[34,35] The fundamental difference between the two models is that Chow and Parish have an enhanced interfacial shear compared with Pohner, and therefore their liquid film is thinner, the heat-transfer coefficient higher, condensation more rapid, and thus quality decreases more quickly than Pohner's.

The model calculations for Fig. 26 are based on an inlet quality of 95%. Lee and Best's inlet condition was approximately 5% quality. The quality decreases of the two-phase mixture in passing through the cooling jackets was 2.2% for jacket 1 and 2.3% for jacket 2 for parabolas 13–15. None of

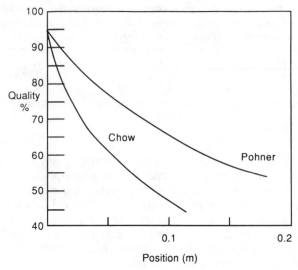

Fig. 26 Comparative plot of condensation models of Pohner[13] and Chow and Parish.[11]

the models is considered to be validated, and Lee and Best's data are questionable. Much more quantitative work must be done.

Dropwise Condensation

There are no quantitative data available on microgravity dropwise condensation. Descriptive information on mercury condensation has been presented in the section on Mercury Condensation Experiments.

Bulk Condensation

There are no quantitative data available on bubbly/slug microgravity condensation.

Design Recommendations

No microgravity condensation model has been sufficiently validated to be considered a reliable design tool. Pohner's model is the only one that has been validated against data of any kind, albeit the data was ground based. Since Pohner's model is more conservative than that of Chow and Parish, the best available approach seems to be to use Pohner's model from the entrance at some quality to a quality of zero and to compute the single-phase liquid heat-transfer coefficient, using the higher of the two at a given axial position.

State of the Art

Summary
This chapter has presented condensation models developed for single-component two-phase flow for several geometries and flow conditions. Various microgravity condensation experiments have been described; however, their one common feature is that quantitative data suitable for model validation have not been collected. Experiments using mercury, Freons, and water have qualitatively demonstrated that stable microgravity condensation can occur in properly designed systems. Pressure recovery due to condensation has been qualitatively observed, but quantitative model validation remains to be carried out. Many experiments are planned to investigate microgravity two-phase condensation.

Planned Experiments
Microgravity condensation technology demonstration experiments are being carried out by NASA in conjunction with many corporations. Grumman has a two-phase thermal bus technology demonstration loop[36,37]; Boeing has a two-phase thermal bus[38]; McDonnell-Douglas the two-phase thermal control flight experiment[39]; and Lockheed the two-phase thermal bus.[40] All of these projects include ground testing, KC-135 tests, and Shuttle experiments.

The Air Force has three microgravity two-phase flow experiments that include quantitative measurement of boiling and condensation processes, as well as system-level two-phase flow instabilities. The Air Force (WL/AWYS) has funded Creare, Inc.[41] to develop and verify design methods for two-phase flow and heat transfer in microgravity. Texas A&M University supplies some flow modeling and KC-135 experiment design and flight support for this project.

The Air Force (WL/AWYS) has funded the Foster Miller Corporation[42] to carry out fundamental microgravity two-phase flow studies. Texas A&M University supplies phenomenological modeling and KC-135 experiment design and flight support.

The Air Force (WL/AWYS) also has an in-house suborbital flight experiment for a boiling/condensing system[43] to be flown in 1990.

Concluding Comments
Many models exist for microgravity condensation phenomena, and several projects are being carried out to validate these models and design methodologies. NASA is supplying an important facet of this technology development by holding microgravity meetings[44,45] to disseminate information about planned activities. A key recommendation of these meetings has been that a coordinated test program be established to generate test data necessary to develop and validate microgravity two-phase flow models. Nevertheless, all interested parties should realize that two-phase flow experiments for Earth-based systems are still being carried out, that the

fundamental nature of two-phase flow is still being explored, and that two-phase flow experiments in microgravity will come into their own when long-term orbital research labs become available.

References

[1]"Industrial Applications of the Microgravity Environment," Space Applications Board Commission on Engineering and Technical Systems, National Research Council, Washington, DC, March 1988.

[2]Mankins, J. C., Hepenstal, A., and Olivieri, J., "Civil Mission Applications of Space Nuclear Power," Jet Propulsion Lab., Pasadena, CA, JPL D-3547, Oct. 1986.

[3]"Assessment of Developments in Space Power," Arthur D. Little Co., Cambridge, MA, Ref. 60847, March 1988.

[4]Nusselt, W., "Die Oberflachen Kondensation des Wasserdampfes," *VDI Zeitschrift*, Vol. 60, 1916, pp. 541–569.

[5]Rohsenow, W. M. and Choi, H. Y. H., *Heat Mass and Momentum Transfer*, Prentice-Hall, Englewood Cliffs, NJ, 1961.

[6]Rohsenow, W. M., "Film Condensation," *Applied Mechanics Reviews*, Vol. 23, May 1970, pp. 487–495.

[7]Gröber, H., Erk, S., and Grigull, U., "Fundamentals of Heat Tranfer," McGraw-Hill, New York, 1961, p. 313.

[8]Keshock, E. G. and Sadeghipour, M. S., "Analytical Comparison of Condensing Flows Inside Tubes Under Earth-Gravity and Space Environments," *Acta Astronautica*, Vol. 10, July–Dec. 1983, pp. 505–511.

[9]Rufer, C. E. and Kezios, S. P., "Analysis of Two Phase One Component Stratified Flow with Condensation," *Journal of Heat Transfer*, Vol. 88, 1966, pp. 265–275.

[10]Bae, S., Malbetsch, J. S., and Rohsenow, W. M., "Refrigerant Forced Convection Condensation Inside Horizontal Tubes," Massachustetts Institute of Technology, Mechanical Engineering Dept., Cambridge, MA, Rept. DSR 72591-71, 1970.

[11]Chow, L. C. and Parish, R. C., "Condensation Heat Transfer in a Microgravity Environment," *Proceedings of the 24th Aerospace Sciences Meeting*, AIAA, New York, 1986.

[12]Henstock, W. H. and Hanratty, T. J., "The Interfacial Drag and the Height of the Wall Layer in Annular Flows," *AIChE Journal*, Vol. 22, 1976, pp. 990–1000.

[13]Pohner, J. A., "A Nusselt-Type Analysis of Steady, Condensing and Evaporating Flows in Circular Tubes," AIAA Paper 86-1328, June 1986.

[14]Goodykoontz J. H., and Dorsch, R. G., "Local Heat Transfer Coefficients for Condensation of Steam in Vertical Downflow within a 5/8 Inch Diameter Tube," NASA TN D-3326, 1966.

[15]Faghri, A. and Chow, L. C., "Annular Condensation Heat Transfer in a Microgravity Environment," AIAA Paper 87-1533, June 1987.

[16]Eastman, R. E., Feldmanis, C. J., Haskin, W. L., and Weaver, K. L., "Two-Phase Fluid Thermal Transport for Spacecraft," Vehicle Equipment Division, Flight Dynamics Lab., AFWAL-TR-84-3028, Wright-Patterson Air Force Base, OH, Oct. 1984.

[17]Collier, J. G., *Convective Boiling and Condensation* (2nd ed.), McGraw-Hill, New York, 1981, pp. 323–328.

[18]Lancet, R. T., Abramson, P., and Forslund, R. P., "The Fluid Mechanics of Condensing Mercury in a Low-Gravity Environment," *Fluid Mechanics and Heat Transfer Under Low Gravity*, edited by H. Cohan and M. Rogers, 1965.

[19]Griffith, P., "Dropwise Condensation (for Water on Various Surfaces)," Mechanical Engineering Dept., Massachusetts Institute of Technology, Cambridge, MA, 1976.

[20]Best, F. R., Kachnik, L., and Lee, D.-J., "Space Microgravity Two Phase Flow Safety Considerations," *Transactions of the American Nuclear Society*, Vol. 52, June 1986.

[21]Reitz, J. G., "Zero Gravity Mercury Condensing Research," *Aerospace Engineering*, TL 501 A 3.26, Vol. 19, Sept. 1960, p. 18.

[22]Albers, J. A. and Macosko, R. P., "Experimental Pressure-Drop Investigation of Nonwetting, Condensing Flow of Mercury Vapor in a Constant-Diameter Tube in 1-G and Zero-Gravity Environments," NASA TN D-2838, June 1965.

[23]Albera, J. A. and Namkoong, D. Jr., "An Experimental Study of the Condensing Characteristics of Mercury Vapor Flowing in Single Tubes," AIAA Paper XX-XXXX, 1965.

[24]Albers, J. A. and Macosko, R. P., "Condensation Pressure Drop of Nonwetting Mercury in a Uniformly Tapered Tube in 1-g and Zero-gravity Environments," NASA TN D-3185, Jan. 1966.

[25]Sturas, J. I., Crabs, C. C., and Gorland, S. H., "Photographic Study of Mercury Droplets Parameters Including Effects of Gravity," NASA TN D-3705, Nov. 1966.

[26]Namkoong, D., Block, H. B., Macosko, R. P., and Crabs, C. C., "Photographic Study of Condensing Mercury Flow in 0- and 1-G Environments," NASA TN D-4023, June 1967.

[27]"Two-Phase Mercury Flow in Zero-Gravity (film)," NASA Lewis Research Center, Cleveland, OH, Technical Information Division (MS 5-5), c 221, 1966.

[28]"Photographic Study of Non-Wetting Condensing Mercury Flow at One and Zero-G," NASA Lewis Research Center, Cleveland, OH, Technical Information Division (MS 5-5), c 251, 1967.

[29]Koestel, A., Gutstein, M., and Wainwright, R., "Study of Wetting and Non-Wetting Mercury Condensing Pressure Drops," NASA TN D-2514, 1964.

[30]Williams, J. L., Keshock, E. G., and Wiggins, C. L., "Development of a Direct Condensing Radiator for Use in a Spacecraft Vapor Compression Refrigeration System," *Transactions of the American Society of Mechanical Engineers*, Vol. 95, Nov. 1973, pp. 1053–1064.

[31]Chen, I.-Y, Dowing, R. S., Parish, R., and Keshock, E., "A Reduced Gravity Flight Experiment: Observed Flow Regimes and Pressure Drops of Vapor and Liquid Flow In Adiabatic Piping," *American Institute of Chemical Engineering Symposium Series*, Vol. 84, No. 263, 1988, pp. 203–216.

[32]"A Study of Two-Phase Flow in a Reduced Gravity Environment," Sundstrand Energy Systems, Sundstrand Corp., Rockford, IL, DRD No. TM-478T, Oct. 16, 1987.

[33]Feldmanis, C. J., "Pressure and Temperature Changes in Closed Loop Forced Convection Boiling and Condensing Processes Under Zero Gravity Conditions," *Institute of Environmental Sciences 1966 Annual Technical Meeting Proceedings*, 1966, pp. 455–461.

[34]Lee, D., "Thermohydraulic and Flow Regime Analysis for Condensing Two Phase Flow in a Microgravity Environment," Ph.D. Thesis, Nuclear Engineering Dept., Texas A&M Univ., College Station, TX, Dec. 1987.

[35]Best, F. R., Lee, D.-J., and Kachnik, L., "KC-135 Reduced Gravity Flow Regime, Boiling and Condensing Water Experiments," Interphase Transport Phenomena Lab. Rept. 89-2, Texas A&M Univ., Nuclear Engineering Dept., College Station, TX, May 1989.

322

F. R. BEST

[36]Edelstein, F. and Brown, R., "Initial Testing of a Two-Phase Thermal Loop," AIAA Paper 86-1296, June 1986.

[37]Edelstein, F., Liandris, M., and Rankin, J. G., "Thermal Test Results of the Two-Phase Themal Bus Technology Demonstration Loop," AIAA Paper 87-1627, June 1987.

[38]Myron, D. L. and Parish, R. C., "Development of a Prototype Two-Phase Thermal Bus System for Space Station," AIAA Paper 87-1628, June 1987.

[39]Grote, M. G., Stark, J. A., Butler, C. D., and McIntosh, R., "Design and Test of a Mechanically Pumped Two-Phase Thermal Control Flight Experiment," AIAA Paper 87-1629, June 1987.

[40]Holmes, H. R., Goepp, J. W., and Hewitt, H. W., "Development of the Lockheed Pumped Two-Phase Thermal Bus," AIAA Paper 87-1626, June 1987.

[41]"Test Plan for Ground Tests of Key Components in a Microgravity Experimental Two Phase Flow System," Creare, TM-1334, March 1989.

[42]Hill, W. S., "Definition of Microgravity Two Phase Flow Behaviors for Spacecraft Design," presented at the NASA Center for Space Power, Texas A&M Univ., College Station, TX, Dec. 15–16, 1988.

[43]"Two Phase Flow in Microgravity Experiment," U.S. Air Force, WL/AWYS, Kirtland Air Force Base, May 1989.

[44]"Microgravity Fluid Management Symposium," NASA CP-2465, Sept. 1986.

[45]Swanson, T. and Henstein, L. U. (eds.), "Workshop on Two Phase Fluid Behavior in a Space Environment," NASA-OAST, Ocean City, MD, June 13–14, 1988.

Author Index

PROGRESS IN ASTRONAUTICS AND AERONAUTICS
SERIES VOLUMES

*1. **Solid Propellant Rocket Research** (1960)
Martin Summerfield
Princeton University

*2. **Liquid Rockets and Propellants** (1960)
Loren E. Bollinger
Ohio State University
Martin Goldsmith
The Rand Corp.
Alexis W. Lemmon Jr.
Battelle Memorial Institute

*3. **Energy Conversion for Space Power** (1961)
Nathan W. Snyder
Institute for Defense Analyses

*4. **Space Power Systems** (1961)
Nathan W. Snyder
Institute for Defense Analyses

*5. **Electrostatic Propulsion** (1961)
David B. Langmuir
Space Technology Laboratories, Inc.
Ernst Stuhlinger
NASA George C. Marshall Space Flight Center
J.M. Sellen Jr.
Space Technology Laboratories, Inc.

*6. **Detonation and Two-Phase Flow** (1962)
S.S. Penner
California Institute of Technology
F.A. Williams
Harvard University

*7. **Hypersonic Flow Research** (1962)
Frederick R. Riddell
AVCO Corp.

*8. **Guidance and Control** (1962)
Robert E. Roberson,
Consultant
James S. Farrior
Lockheed Missiles and Space Co.

*9. **Electric Propulsion Development** (1963)
Ernst Stuhlinger
NASA George C. Marshall Space Flight Center

*10. **Technology of Lunar Exploration** (1963)
Clifford I. Cummings
Harold R. Lawrence
Jet Propulsion Laboratory

*11. **Power Systems for Space Flight** (1963)
Morris A. Zipkin
Russell N. Edwards
General Electric Co.

*12. **Ionization in High-Temperature Gases** (1963)
Kurt E. Shuler, Editor
National Bureau of Standards
John B. Fenn,
Associate Editor
Princeton University

*13. **Guidance and Control – II** (1964)
Robert C. Langford
General Precision Inc.
Charles J. Mundo
Institute of Naval Studies

*14. **Celestial Mechanics and Astrodynamics** (1964)
Victor G. Szebehely
Yale University Observatory

*15. **Heterogeneous Combustion** (1964)
Hans G. Wolfhard
Institute for Defense Analyses
Irvin Glassman
Princeton University
Leon Green Jr.
Air Force Systems Command

16. **Space Power Systems Engineering** (1966)
George C. Szego
Institute for Defense Analyses
J. Edward Taylor
TRW Inc.

17. **Methods in Astrodynamics and Celestial Mechanics** (1966)
Raynor L. Duncombe
U.S. Naval Observatory
Victor G. Szebehely
Yale University Observatory

18. **Thermophysics and Temperature Control of Spacecraft and Entry Vehicles** (1966)
Gerhard B. Heller
NASA George C. Marshall Space Flight Center

*19. **Communication Satellite Systems Technology** (1966)
Richard B. Marsten
Radio Corporation of America

---

*Out of print.

(Other Volumes are planned.)